Construction Materials and Testing

AUTHOR

Dr. Vanita Aggarwal
Professor and Head
Department of Civil Engineering
Maharishi Markandeshwar (Deemed to be University)
Mullana (Ambala), Haryana

REVIEWER

Dr. A.D. Prasad
Assistant Professor
Department of Civil Engineering
National Institute of Technology
Raipur, Chhattisgarh

KHANNA BOOK PUBLISHING CO. (P) Ltd.
4C/4344, Ansari Road, Darya Ganj,
New Delhi, India

All India Council for Technical Education
Nelson Mandela Marg, Vasant Kunj,
New Delhi, India

Dear Readers,

To prevent the piracy, this book is secured with HIGH SECURITY HOLOGRAM on the front title cover. In case you don't find the hologram on the front cover title, please write us to at contact@khannabooks.com or whatsapp us at +91-99109 09320 and avail special gift voucher for yourself.

Specimen of Hologram on front Cover title:

Moreover, there is a SPECIAL DISCOUNT COUPON for you with EVERY HOLOGRAM.

How to avail this SPECIAL DISCOUNT:

Step 1: Scratch the hologram

Step 2: Under the scratch area, your "coupon code" is available

Step 3: Logon to www.khannabooks.com

Step 4: Use your "coupon code" in the shopping cart and get your copy at a special discount

Step 5: Enjoy your reading!

ISBN: *978-93-5538-333-4*

Book Code: DIP449EN

Construction Materials and Testing

by Dr. Vanita Aggarwal

First Edition: 2026

Originally Published by:
All India Council for Technical Education, Ministry of Education, GOI

Exclusively Published & Distributed by:
Khanna Book Publishing Co. (P) Ltd.
4C/4344, Ansari Road, Daryaganj,
New Delhi (INDIA)

Visit us at: **www.khannabooks.com**
Write us at: **contact@khannabooks.com**

To view complete list of books, Please scan the QR Code: KPH

Printed in Bharat.

प्रो. टी. जी. सीताराम
अध्यक्ष
Prof. T. G. Sitharam
Chairman

सत्यमेव जयते

आज़ादी का अमृत महोत्सव

अखिल भारतीय तकनीकी शिक्षा परिषद्
(भारत सरकार का एक सांविधिक निकाय)
(शिक्षा मंत्रालय, भारत सरकार)
नेल्सन मंडेला मार्ग, वसंत कुंज, नई दिल्ली–110070
दूरभाष : 011–26131498
ई–मेल : chairman@aicte-india.org

ALL INDIA COUNCIL FOR TECHNICAL EDUCATION
(A STATUTORY BODY OF THE GOVT. OF INDIA)
(Ministry of Education, Govt. of India)
Nelson Mandela Marg, Vasant Kunj, New Delhi-110070
Phone : 011-26131498
E-mail : chairman@aicte-india.org

FOREWORD

Engineers are the backbone of any modern society. They are the ones responsible for the marvels as well as the improved quality of life across the world. Engineers have driven humanity towards greater heights in a more evolved and unprecedented manner.

The All India Council for Technical Education (AICTE), have spared no efforts towards the strengthening of the technical education in the country. AICTE is always committed towards promoting quality Technical Education to make India a modern developed nation emphasizing on the overall welfare of mankind.

An array of initiatives has been taken by AICTE in last decade which have been accelerated now by the National Education Policy (NEP) 2020. The implementation of NEP under the visionary leadership of Hon'ble Prime Minister of India envisages the provision for education in regional languages to all, thereby ensuring that every graduate becomes competent enough and is in a position to contribute towards the national growth and development through innovation & entrepreneurship.

One of the spheres where AICTE had been relentlessly working since past couple of years is providing high quality original technical contents at Under Graduate & Diploma level prepared and translated by eminent educators in various Indian languages to its aspirants. For students pursuing 2nd year of their Engineering education, AICTE has identified 87 books, which shall be translated into 12 Indian languages - Hindi, Tamil, Gujarati, Odia, Bengali, Kannada, Urdu, Punjabi, Telugu, Marathi, Assamese & Malayalam. In addition to the English medium, books in different Indian Languages are going to support the students to understand the concepts in their respective mother tongue.

On behalf of AICTE, I express sincere gratitude to all distinguished authors, reviewers and translators from the renowned institutions of high repute for their admirable contribution in a record span of time.

AICTE is confident that these outcomes based original contents shall help aspirants to master the subject with comprehension and greater ease.

(Prof. T. G. Sitharam)

प्रो. टी. जी. सीताराम
अध्यक्ष
Prof. T. G. Sitharam
Chairman

अखिल भारतीय तकनीकी शिक्षा परिषद
ALL INDIA COUNCIL FOR TECHNICAL EDUCATION
(A Statutory Body of the Govt. of India)
Ministry of Education, Govt. of India
Nelson Mandela Marg, Vasant Kunj, New Delhi-110070
Phone : 011-26131498
E-mail : chairman@aicte-india.org

FOREWORD

Engineers are the backbone of any modern society. They are the ones responsible for the marvels as well as the improved quality of life across the world. Engineers have driven humanity towards greater heights in a more evolved and unprecedented manner.

The All India Council for Technical Education (AICTE), have spared no efforts towards the strengthening of the technical education in the country. AICTE is always committed towards promoting quality Technical Education to make India a modern developed nation emphasizing on the overall welfare of mankind.

An array of initiatives has been taken by AICTE in last decade which have been accelerate now by the National Education Policy (NEP) 2020. The implementation of NEP under the visionary leadership of Hon'ble Prime Minister of India envisages the provision for education in regional languages to all, thereby ensuring that every graduate becomes competent enough and is in a position to contribute towards the national growth and development through innovation & entrepreneurship.

One of the spheres where AICTE had been relentlessly working since past couple of years is providing high quality original technical contents at Undergraduate & Diploma level prepared and translated by eminent educators in various Indian languages to its aspirants. For students pursuing 2nd year of their Engineering education, AICTE has identified 37 books, which shall be translated into 12 Indian languages - Hindi, Tamil, Gujarati, Odia, Bengali, Kannada, Urdu, Punjabi, Telugu, Marathi, Assamese & Malayalam. In addition to the English medium, books in different Indian Languages are going to support the students to understand the concepts in their respective mother tongue.

On behalf of AICTE, I express sincere gratitude to all distinguished authors, reviewers and translators from the renowned institutions of high repute for their admirable contribution in a record span of time.

AICTE is confident that these outcomes based original contents shall help aspirants to master the subject with comprehension and greater ease.

(Prof. T. G. Sitharam)

ACKNOWLEDGEMENT

The author is grateful to the authorities of AICTE, particularly Prof. T. G. Sitharam, Chairman; Dr. Abhay Jere, Vice-Chairman; Prof. Rajive Kumar, Member-Secretary; Dr. Ramesh Unnikrishnan, Advisor-II and Dr. Sunil Luthra, Director, Training and Learning Bureau, for their planning to publish the books on Civil Engineering. I sincerely acknowledge the valuable contributions of the reviewer of the book Prof. A.D. Prasad, Civil Engineering, NIT Raipur for his valuable inputs to enhance the quality and better shape the book to make it student friendly.

My special thanks are due to MM (DU), Mullana officials for providing constant moral support, and encouragement. I want to thank my supporting staff Mr. Harpreet to help me draft the manuscript. I also thank all my students and fellow faculty members for inspiring me during my years of service, learnings from which have proved valuable in publishing the contents of this book.

I would like to express my deepest thanks to my husband Dr. Viney Aggarwal whose motivation has been with me throughout this noble journey. I wish to express my gratitude to my children, Priysha and Manit for their unconditional love, care, and strength which allowed me to finish our book successfully.

Last, but definitely not the least, I would like to dedicate this book to my father, Late Sh. Kalwant Rai whose teachings abled me to reach this position today.

This book is an outcome of various suggestions of AICTE members, experts and authors who shared their opinions and thoughts to further develop the engineering education in our country. Acknowledgements are due to the contributors and experts in this field whose published books, review articles, papers, photographs, footnotes, references and other valuable information enriched me at the time of writing the book. I hope this book will help civil engineering students understand the subject easily and conveniently.

Dr. Vanita Aggarwal

PREFACE

The book titled "Construction Materials and Testing" is an outcome of the extensive experience of my teaching of Building Construction Materials and Drawing, and Concrete Technology courses and allied research work carried out on construction materials. The purpose of writing this book is to provide deep knowledge of construction materials to the students and professionals in Civil Engineering. Keeping in mind the purpose of wide coverage as well as to provide essential supplementary information, the topics recommended by AICTE have been included, in a very systematic and orderly manner throughout the book. Efforts have been made to explain the fundamental concepts of the subject in the simplest possible way.

During the process of preparation of the manuscript, various standard textbooks have been referred thoroughly and accordingly the sections like multiple choice questions, short and long answer questions and practical exercises have been framed. While preparing the different sections, emphasis has also been laid on definitions and principles of selection of appropriate material for the desired application. As the subject is more about real life application of construction materials, many examples of practical projects have been discussed in "Know More" section. The book covers all types of details of building construction materials required for a Civil Engineer and these have been presented in a very logical and systematic manner.

Apart from illustrations, examples and exercises as required, the book has been enriched with numerous QR codes describing additional details for the curious minds. The book is written in a way that it ignites the inquisitive approach of its readers. It explains the fundamental concepts, material characteristics and related laboratory experiments on natural, artificial, waste-based and specially processed building construction materials. This will enable students to contribute creatively to infrastructure projects throughout the whole construction process, from the foundation to the finished structure.

It is expected that the book is easy to comprehend and will inspire civil engineering fraternity at large to learn and apply suitable construction material at all times. This will surely contribute to the development of a solid foundation of the subject. All beneficial comments and suggestions that will contribute to the improvement of future editions of the book are always welcome and may be sent to aggarwal_vanita@rediffmail.com. It is an immense pleasure to publish this book for the benefit of teachers, students, and civil engineering professionals.

Dr. Vanita Aggarwal

OUTCOME BASED EDUCATION

For the implementation of an outcome based education the first requirement is to develop an outcome based curriculum and incorporate an outcome based assessment in the education system. By going through outcome based assessments, evaluators will be able to evaluate whether the students have achieved the outlined standard, specific and measurable outcomes. With the proper incorporation of outcome based education there will be a definite commitment to achieve a minimum standard for all learners without giving up at any level. At the end of the programme running with the aid of outcome based education, a student will be able to arrive at the following outcomes:

Programme Outcomes (POs) are statements that describe what students are expected to know and be able to do upon graduating from the program. These relate to the skills, knowledge, analytical ability attitude and behaviour that students acquire through the program. The POs essentially indicate what the students can do from subject-wise knowledge acquired by them during the program. As such, POs define the professional profile of an engineering diploma graduate.

National Board of Accreditation (NBA) has defined the following seven POs for an Engineering diploma graduate:

PO1. Basic and Discipline specific knowledge: Apply knowledge of basic mathematics, science and engineering fundamentals and engineering specialization to solve the engineering problems.

PO2. Problem analysis: Identify and analyses well-defined engineering problems using codified standard methods.

PO3. Design/development of solutions: Design solutions for well-defined technical problems and assist with the design of systems components or processes to meet specified needs.

PO4. Engineering Tools, Experimentation and Testing: Apply modern engineering tools and appropriate technique to conduct standard tests and measurements.

PO5. Engineering practices for society, sustainability and environment: Apply appropriate technology in context of society, sustainability, environment and ethical practices.

PO6. Project Management: Use engineering management principles individually, as a team member or a leader to manage projects and effectively communicate about well-defined engineering activities.

PO7. Life-long learning: Ability to analyse individual needs and engage in updating in the context of technological changes.

COURSE OUTCOMES

After completion of the course the students will be able to:

CO-1: Identify relevant construction materials.

CO-2: Identify relevant natural construction materials.

CO-3: Select relevant artificial construction materials.

CO-4: Select relevant special type of construction materials.

CO-5: Identify and use of processed construction materials

Mapping of Course Outcomes with Programme Outcomes to be done according to the matrix given below:

Course Outcomes	Expected Mapping with Programme Outcomes (1- Weak Correlation; 2- Medium Correlation; 3- Strong Correlation)						
	PO-1	PO-2	PO-3	PO-4	PO-5	PO-6	PO-7
CO-1	3	1	2	2	3	4	1
CO-2	2	1	3	3	3	2	1
CO-3	2	1	2	3	3	2	1
CO-4	1	2	3	3	3	1	2
CO-5	1	2	2	2	3	3	2

GUIDELINES FOR TEACHERS

To implement Outcome Based Education (OBE) knowledge level and skill set of the students should be enhanced. Teachers should take a major responsibility for the proper implementation of OBE. Some of the responsibilities (not limited to) for the teachers in OBE system may be as follows:

- Within reasonable constraint, they should manoeuvre time to the best advantage of all students.
- They should assess the students only upon certain defined criterion without considering any other potential ineligibility to discriminate them.
- They should try to grow the learning abilities of the students to a certain level before they leave the institute.
- They should try to ensure that all the students are equipped with the quality knowledge as well as competence after they finish their education.
- They should *always encourage the students to develop* their ultimate performance capabilities.
- They should facilitate and encourage group work and team work to consolidate newer approach.
- They should follow Blooms taxonomy in every part of the assessment.

Bloom's Taxonomy

Level	Teacher should Check	Student should be able to	Possible Mode of Assessment
Create	Students ability to create	Design or Create	Mini project
Evaluate	Students ability to justify	Argue or Defend	Assignment
Analyse	Students ability to distinguish	Differentiate or Distinguish	Project/Lab Methodology
Apply	Students ability to use information	Operate or Demonstrate	Technical Presentation/ Demonstration
Understand	Students ability to explain the ideas	Explain or Classify	Presentation/Seminar
Remember	Students ability to recall (or remember)	Define or Recall	Quiz

GUIDELINES FOR STUDENTS

Students should take equal responsibility for implementing the OBE. Some of the responsibilities (not limited to) for the students in OBE system are as follows:

- Students should be well aware of each UO before the start of a unit in each and every course.
- Students should be well aware of each CO before the start of the course.
- Students should be well aware of each PO before the start of the programme.
- Students should think critically and reasonably with proper reflection and action.
- Learning of the students should be connected and integrated with practical and real life consequences.
- Students should be well aware of their competency at every level of OBE.

ABBREVIATIONS AND SYMBOLS

List of Abbreviations

General Terms			
Abbreviations	**Full form**	**Abbreviations**	**Full form**
ACB	Aerated Concrete Blocks	NaOH	Sodium Hydroxide
Al_2O_3	Alumina	OPC	Ordinary Portland Cement
B_2O_3	Boron Oxide	POP	Plaster of Paris
BIS	Bureau of Indian Standards	PPC	Portland Pozzolana Cement
CaO	Calcium Oxide	PPF	Polypropylene Fibers
DPC	Damp Proof Coarse	PUR	Polyurethane Foam
Fe_2O_3	Iron Oxide	PVC	Poly-Vinyl Chloride
FRC	Fibre Reinforced Concrete	RCC	Reinforced Cement Concrete
GGBS	Ground Granulated Blast Furnace Slag	RHC	Rapid Hardening Cement
GRC	Glass fibres Reinforced Concrete	SCBA	Sugar Cane Bagasse Ash
MgO	Magnesium Oxide	SiO_2	Silica
Na_2SiO_3	Sodium Silicate	SO_3	Sulphur Oxide

LIST OF FIGURES

LIST OF TABLES

CONTENTS

1 Overview of Construction Materials

UNIT SPECIFICS

This unit has discussed the following aspects:

- Knowing about various construction materials.
- Classification of construction materials.
- Applications of building materials in different fields of construction like Transportation Engineering, Environmental Engineering, Building Structures, Irrigation Engineering;
- Identification of construction materials in their application.

The practical aspects also taken into consideration to identify various available construction materials in laboratory on the basis of their sources are discussed in this chapter with the aim to generate creativity and curiosity amongst students.

Besides a number of multiple choice questions, one word answer-questions, short and long answer type questions are given following bloom's taxonomy at the end of the unit. Assignments may be framed using these questions by teachers. A list of references and suggested readings are stated in the unit for gaining more knowledge on construction materials. Further, it is noteworthy here that some QR codes have been given in the unit which can be scanned for getting deeper knowledge on the topic.

After discussion of practical in identification of various construction materials, there is a 'Know more' section. This section has been prepared with great care to provide supplemental information on building construction materials for better understanding of the chapter.

RATIONALE

This introductory unit on construction materials helps the students to get primary knowledge about various construction materials available on the earth either naturally or as man-made. It discusses the classification of varied types of building construction materials in the basis of various parameters.

Understanding of classifications is important for a civil graduate to select a proper material for a specified application. The use of construction materials is not limited to any one field of infrastructure but it goes into their applications in housing projects, industrial projects, irrigation structures, reservoirs, transportation sector, treatment plants for water and waste water etc,. The selection of proper materials for a specific purpose is very important in construction industry to ensure cost effective and efficient structure which can serve the purpose for which it is constructed. Further, due importance must be given to the effect of chosen construction materials on environment. As most of the construction materials presently being used, are causing one or the other harmful effect on environment one needs to look for some sustainable alternate construction materials which are eco-friendly. This unit reinforces the basic knowledge on construction materials and helps in identifying right construction materials for the given project in an eco-friendly and economical way.

PRE-REQUISITE

Nil

UNIT OUTCOMES

List of outcomes of this unit is as follows:

U1-O1: Understand the scope of construction materials.

U1-O2: Classify construction materials.

U1-O3: Select appropriate materials in different fields of construction.

U1-O4: Select construction materials on the basis of strength and durability.

U1-O5: Selection of construction materials on the basis of economy and sustainability.

Unit-1 Outcomes	EXPECTED MAPPING WITH COURSE OUTCOMES (1- Weak Correlation; 2- Medium Correlation; 3- Strong Correlation)				
	CO-1	CO-2	CO-3	CO-4	CO-5
U1-O1	1	3	2	1	1
U1-O2	3	3	2	3	1
U1-O3	3	3	2	3	1
U1-O4	3	2	2	3	2
U1-O5	3	1	1	1	3

1.1 SCOPE OF CONSTRUCTION MATERIALS

Construction materials are important for construction of new infrastructures of any kind. The scope of construction materials includes, but not restricted to, use of construction materials in building structures, transportation sector i.e. highways, railways and airways, irrigation structure, hydraulic structures, treatment plants etc. Besides, new construction, construction materials are of great importance for maintenance and repair work of infrastructure. High rise building, chimneys, silos, bridges, flyovers, dams, weir, warehouses, tunnels, pavements, railway tracks, water retaining structures, piers, embankments, sewers, treatment plants are few examples of high order construction projects. Each construction project is constructed to serve a specific purpose i.e. its serviceability and to serve for a specific life span i.e. its durability 'Selection of right construction materials for the right project' defines the expertise of an engineer.

1.1.1 Construction Material in Building Construction

The most common use of construction materials is in building construction. Building construction includes small houses to high rise buildings, residential to commercial spaces, hospitals, academic institutions, hotels to restaurants, cinema halls to amusement parks, industry to public buildings. Building construction is not limited to construction of only new structures, rather, repairs and maintenance during the service life span of a structure by construction materials is also very important. Some of commonly used construction materials for building construction are– Cement, Reinforcement (Steel bars), bricks, mortar, timber, stones, polymer, ceramics, glass, rammed earth.

1.1.2 Construction Material in Transportation Engineering

Transportation engineering includes means of commuting for people and goods. Highways, Railways and Airways are three major types of projects built under Transportation Engineering. Transportation engineering has a great impact on the countries' economy. In fact, a country is known by the infrastructure and transport facilities it holds.

A wide range of materials are used in construction of pavements and roads. Those include soils (natural or processed), aggregates (fine or coarse – natural or processed), binders like bituminous materials, lime or cement and certain admixtures targeted to achieve some specific property at site or to improve overall performance of the roads under traffic.

The basic component of the foundation sub-grade or even road is soil as all the loads from any type of construction or traffic ultimately transmit on soil. Stone aggregates, also called mineral aggregates, are used as sub-bases, granular bases in bituminous roads (flexible pavements) as well as in cement concrete pavements (rigid pavements). Use of aggregates makes the pavement relatively cheaper. A highway engineer needs to give top priority to selection of appropriate type of aggregates giving due emphasis on their properties and characteristics.

The use of bitumen in paving roads is well known to all but it picked up only in nineteenth century. Bitumen is an organic compound which is either taken directly from nature or can be obtained during the distillation of petroleum. The residue to refining industry for crude oil is bitumen and it provides superior riding surface. Another binder used in pavement is cement concrete which provides more durable pavements but are costlier than bituminous ones.

1.1.3 Construction Material in Environmental Engineering

The biggest challenge faced due to the use of construction materials is their impact on environment. Impact includes challenges related to conservation of natural resources being exhausted, pollutants being produced in the processes of manufacturing and usage of construction material and disposal of waste materials produced at construction sites. The development of a sustainable environment and durable infrastructure is desired in the modern times of advanced technology. The construction materials used in various projects are exposed to a wide range of environmental conditions and hence it is important to study the effect of environment on various construction materials. The study of construction materials for environment engineering projects, not only includes use of construction materials but also looks up for environmental protection and sustainable development. The most popular construction material, concrete is also not environmental friendly and an engineer should target to choose a suitable material which does not cause or has minimum harm to environment.

Besides, water and wastewater treatment plants need to be built in residential as well as industrial projects. Construction materials used in building of these treatments plants and need to be assessed for their behaviour towards chemically active environment.

1.1.4 Construction Material in Irrigation Engineering

Irrigation is the science of planning an efficient, economical and artificial water supply system to soil for raising crops. The basic components of an irrigation system consist of an intake structures, pumping systems, conveyance systems, distribution systems and drainage system. For ensuring a regular water supply for agriculture, water is stored in one or the other reservoir like dams, transported to fields through constructed channels like canals, distributed to fields through pipes/cluster, regulated using checks and structures like turnouts. All these structures are required to be constructed by material which can resist erosion due to high velocity water flow and also has minimal seepage loss. Lining to these structures protect the sides and bed of the structure. These can be constructed using various materials like compacted earth, cement concrete, bricks, stones/boulders or even plastics. Concrete is the most popular material for lining, however, it is the most expensive option.

An engineer must select an appropriate non-erodible material for construction of water retaining and supply structures to ensure longevity and durability of the project. Besides, irrigation engineering, choice of construction material is equally significant for other hydraulic structures like water tanks, dams, weirs etc.

1.2 SELECTION OF MATERIALS

In construction industry, selection of appropriate material for specific purpose is very important. In history, there are examples where failures have occurred due to poor material selection. If for a particular application, good tensile strength is required, then one must choose a material with sufficient tensile strength i.e. metals rather than brittle material like bricks or concrete. Similarly, for outdoor applications, materials should be chosen giving due importance to environmental factors. A right material at right place ensures the serviceability of the structures and also tends to increase the life span of the structure. As construction activities are very tedious - involves a lot of finances, energy and time, a minimum maintenance requiring structure with long life span is always desired. Therefore, material selection is of utmost importance for long term success of a construction project.

1.2.1 Factors Affecting Selection of Material

A large number of factors need to be taken in to account for selection of appropriate material for construction purpose. These includes:

- **Strength:** The strength (compressive strength, tensile strength or flexural strength) required for a specific application is different and hence is the foremost property taken into consideration for selection of materials. It is worth noting here that *concrete is strong in compression but weak in tension whereas, steel is strong in tension as well as compression.*
- **Durability:** The span, for which a material will be able to serve the purpose for which it is intended, is defined as its durability. For material selection in construction industry, durability of material is an important parameter. The durability of a product is defined in terms of its properties like mechanical properties, chemical composition, electrical and thermal properties, flexibility, self weight etc.
- **Eco-friendliness:** Most of the artificial construction materials involves energy intensive activities in their production and produce harmful by product or gases which cause adverse effects on

environment. Cement industry is one such example. The material selection must be based on the environmental effect of the chosen materials. Alternative construction material available in nature or which are eco-friendly needs to be promoted in today's era when the condition of environment is alarming. Earthen construction with use of bamboo is an example of eco-friendly construction.

- **Economy:** As construction is a costly industry, the cost of material affects the overall cost of the project. The initial and the maintenance cost of the materials are taken in consideration for selection of appropriate material.

Besides these factors, many other parameters like aesthetic beauty of the material (especially for outdoor projects desiring natural looks), availability of the material and ease of using, affect selection of material. The actual performance of different materials under different loading and environmental conditions as available in historical examples must also be considered while doing material selection for construction activities.

1.2.2 Selection Materials on the Basis of Strength

The strength of a material is the foremost property for its selection. Strength of a material refers to its ability to resist applied loads or force and to retain its original shape and size. Strength of a material depends on its mechanical properties like:

- **Elasticity:** It is the ability of a material to regain its shape and size on removal of stresses/forces applied on it.
- **Plasticity:** It is the ability of being moulded into desired shape.
- **Ductility:** It is the ability of a material to change its shape, like drawing into wires or threads, without losing strength or without breaking.
- **Brittleness:** The material which is hard but is liable to break easily is called brittle.
- **Malleability:** It is the ability to be shaped into thin sheets by hammering or forging.
- **Toughness:** It is the resistance of a material to be broken into two pieces by a crack, called fracture running across it.
- **Hardness:** The ability of a material to resist deformation or indentation on its surface is called hardness.

The combined performance of a material on these parameters defines the strength of a material. The strength of a material is accessed in following types:

- **Tensile Strength:** The ability of materials to resist breaking under tension (length wise pull stress) is called its tensile strength.
- **Compressive Strength:** The resistance of a material to breaking under compressive forces (loads tending to reduce size) is called its compressive strength.
- **Shearing Strength:** The resistance of a material to with stand shearing load i.e. the forces that tend to produce sliding failure of a material along a plane that is parallel to the direction of force is called shearing strength.

It is worth noting here that steel is a material which has good compressive as well as tensile strength. However, concrete is strong against compressive forces but weak under tensile loading.

1.2.3 Selection of Materials on the Basis of Durability

Durability of a material refers to remain serviceable during the useful life span without damage or unexpected maintenance. The long term performance of a material is generally accepted depending on the previous experience of designer for that material. Sometimes, re-usable property of a material is also considered under its durability. As construction projects extensity involve huge money, time and energy, durability of construction materials becomes very important. Concrete, stone and steel are considered superior to timber, glass and bricks in terms of durability or service life for their structures.

1.2.4 Selection of Materials on their Eco-Friendly Nature

The eco-friendliness of a construction material is defined as its ability to ensure optimal use of natural resources, produce minimum waste and ensure safety towards people and protection of environment. Further, an eco-friendly material is reusable or recyclable and durable. Some of the common eco-friendly construction material includes– Earth (soil), Bamboo, Recycled Steel, Fibres etc. However, the most popular construction material, concrete is not eco-friendly as the production of one the basic material of concrete i.e. cement is an energy extensive process which required huge fuels and also produce large amount of heat and green house gases. Also concrete is non-reusable material. The use of many waste materials like fly ash, slag, recycled aggregates, demolition wastes, industrial waste like metakaolin etc. in concrete are being tried to make concrete a greener material.

1.2.5 Selection of Materials on Basis of Economy

The cost of a material is calculated in terms of its initial cost, handling cost, maintenance cost and re-sale cost. The construction material should be selected considering cost of a material, as more than 50% of the cost of an infrastructure project is its materials. Conventional concrete is a costly material but use of waste materials in concrete can make it economical to some extent. Steel as construction material has higher initial cost but its maintenance cost is less. Bamboo is economical than timber. Brick masonry structures are cheaper in their overall cost than stone masonry but the durability of stone masonry structure is more than brick masonry ones. Recently, some portion of low cost materials and wastes are being mixed in an appropriate proportion to make cement concrete keeping in view economic and sustainability aspects.

A civil engineer must consider all these factors for choice of right construction material for right purpose in a specified project. No construction project is made by a single construction material. Varied construction materials are selected for specific purpose in a construction project. For example– Concrete is used in flexural and compression structural members like beam, slabs, columns, foundation etc. Steel bars are used in these members as reinforcement to bear tensile and additional compressive strength. Steel is popular material in construction due to its versatility, sustainability and flexibility. Non load bearing member like walls, facia, parapets are made up of brick masonry. Metal is used in framing and roofing. Stone masonry is commonly used for exterior walls and fire places. Timbers and glass are used to make doors and windows considering ventilation and sunlight

1.3 CLASSIFICATION OF MATERIALS

A broad classification of construction materials is as follows:

- Natural Materials
- Artificial Material
- Special Construction Material
- Finishing Materials
- Recycled Materials

1.3.1 Natural Building Materials

The materials that are present naturally in environment and can be used for construction purpose are called natural building materials.

Some of the examples of natural materials are sand, clay, gravels, rocks and metals etc. Sand is present naturally on the beds of the river. Rocks are naturally present big boulders. Stones in different sizes are obtained from quarry or through rock blasting.

1.3.1.1 Rocks

Rocks are big boulders that are formed naturally at the crust of the earth. Many historical ancient structures are constructed with rocks e.g. Egyptian Pyramids. Rocks are strong in compression and can support large loads. Mortar is used to hold rocks together.

1.3.1.2 Clay

Clay and mud are natural construction materials which are popular even in modern construction. Clay has a high thermal mass which helps it to keep cool during the summers and warm in winters. A constant temperature can be maintained in clayed structures. It reduces the energy consumption of the building in running air conditioners and heaters.

1.3.1.3 Sand

Sand is formed at the river bed by decomposition of sandstone due to weathering action; Sand is used as a filling material in concrete and is responsible for volume increase in the mix, making it compact and economical.

1.3.1.4 Timber

Wood used for engineering purpose is called timber. In building construction, timber is used to make doors, windows and furniture. Besides, timber is used as material for shuttering and framework in reinforced concrete structures. Sometimes it is used as roofing and flooring material also.

1.3.2 Artificial Building Materials

The man made materials which are used for various construction purposes are called Artificial Building Materials. Concrete, bricks, steel, glass, artificial sand, plastics are some of the popular artificial materials used in construction.

1.3.2.1 Concrete

Concrete is the most popular and versatile building material. Concrete is a mixture of cement, sand, aggregates in wet state. Concrete has good compressive strength, but it is weak in tension. Steel bars are provided in concrete to take care of the tensile stresses and it is called Reinforced Cement Concrete (RCC). Concrete is used in construction of structural members like beams, slabs, columns and foundation in buildings.

1.3.2.2 Bricks

The second most popular artificial building material is brick. It is widely in construction of walls. Bricks are composed of mainly clay involving the process of burning at very high temperature in kiln. Bricks have good compressive strength. Bricks are available in market in standard sizes and varied colours and hence are used for desired appearance of the structure.

1.3.2.3 Steel

Steel is used in many forms in construction activities. These include reinforcing bars, trusses, angle-sections etc. Steel has good compressive as well as tensile strength. Some treatments are carried out on steel to make it resistant to corrosion. Steel has good tensile strength and durability.

1.3.2.4 Glass

Glass is a non crystalline amorphous material. It is used in construction where sunlight is required. Window panels are mostly provided with glass. Glass in toughened form is also used for partition walls and decorative looks to enhance the aesthetic appearance of the building.

1.3.2.5 Artificial Sand

Artificial sand, also known as manufactured sand, is made by mechanical cursing of stones. The use of artificial sand is increasing day by day as there is a shortage of river sand. Artificial sand is used in mortar and concrete.

1.3.2.6 Plastic

Plastics is used in construction in the form of sheets and pipes etc. Plastic is an economical alternative and is easy to handle.

1.3.3 Special Construction Materials

The construction materials which are manufactured specially to possess some special desired properties and are used for specific purposes fall under this category. These include Poly-Vinyl Chloride (PVC) materials, Polypropylene fibbers (PPF), Emulsion and paints etc.

1.3.3.1 Poly-Vinyl Chloride (PVC)

PVC is an economical construction material which has very good insulating properties, good water resistance and good fire resistance. PVC sheets are used to make electrical panels and boards in buildings. It is a light weight material and is easy to use and assemble. PVC pipes are used as water and waste water carrying pipes. PVC sheets are used as flooring, roofing and wall panelling to give decorative look to the structure.

1.3.3.2 Polypropylene Fibres (PPF)

A polypropylene fibre has good compressive and tensile strength. These are used in reinforced concrete as a partial replacement to steel to take up tensile stresses.

1.3.3.3 Emulsions and Paints

Paints of various types are available in the market. Emulsions are oil based paints which are prepared by dispersion of bitumen in water in presence of some stabilizers like oleic acid. Emulsions provide a smooth water proof coating which ensure a good finish also. Bitumen is provided as a coating at Damp Proof Coarse (DPC) level in the building.

1.3.4 Finishing Materials

The construction materials which are used to provide a smooth finish on the surfaces are called finishing materials. Finishing materials includes lime plaster and Plaster of Paris (POP) etc.

1.3.4.1 Lime Plaster

The surfaces of brick walls are covered with a layer of lime plaster to act as a smooth surface. Lime mortar is prepared by mixing sand and lime in equal proportions. Lime plaster has a good binding property to brick masonry and concrete.

1.3.4.2 Plaster of Paris (POP)

Gypsum when heated upto a certain temperature results in a fine powder which is called Plaster of Paris. During heating the water of crystallization gets removed from gypsum. Plaster of Paris is a popular material of plastering. POP sets immediately when water is added to it. This form of plaster is used for interior architectural works as sharp edges and corners can be achieved using POP.

1.3.5 Recycled Materials

The waste materials from various industries, agriculture and even waste from demolished structures are used as recycled materials in construction. These include fly Ash (a waste from thermal power plants), timber creek, agricultural by-products like bagasse, stubble, recycled aggregates etc. Timber creek is prepared by blending saw mill dust, sand, cement and some binder. It can be moulded to prepare blocks, bricks, cubes etc. Fly ash has excellent cementitious properties and can be used as partial replacement to cement and sand. Agricultural wastes by products are added as supplement material to concrete to act as mineral admixture to enhance its properties. Recycled aggregates are generally used as a filler material under flooring. All recycled materials are used in construction have an advantage of economy and energy.

UNIT SUMMARY

- Sectors for use of Construction Material
  - Building Construction
  - Transportation Engineering
  - Environmental Engineering
  - Irrigation and Water Resources Engineering
- Properties desired in Construction Materials
  - Strength
  - Durability
  - Serviceability
  - Pleasing look (Appearance)
- Factors affecting selection of Building Construction Materials
  - Load Carrying Capacity (Strength)
  - Cost
  - Eco-Friendliness
  - Life Span (Durability)
  - Ease of Availability
  - Appearance
- Classification of Construction Materials
  - Natural Materials e.g. Stone, Sand, Clay
  - Artificial Materials e.g. Concrete, Brick, Glass

  - Special Materials e.g. PVC, PPF
  - Finishing Materials e.g. Lime Plaster, POP
  - Recycled Materials e.g. Timber Creek, Recycled Aggregates
- Commonly used Construction Materials for Buildings
  - Mud and Clay
  - Bricks and Blocks
  - Sand
  - Aggregate
  - Stone
  - Concrete
  - Timber
  - Steel

EXERCISES

Multiple Choice Questions

1. Stones from natural rocks are taken out by:
(a) Quarrying (b) Weathering (c) Dressing (d) Cladding
2. Which of the following is natural construction material?
(a) Concrete (b) Sand (c) Cement (d) Glass
3. The resistance to fail by sliding is called
(a) Compressive Strength (b) Tensile Strength
(c) Shearing Strength (d) None of the above
4. The property to regain shape and size on removal of load is called
(a) Elasticity (b) Plasticity (c) Hardness (d) Toughness
5. Which material is preferred for making fireplaces?
(a) Bricks (b) Stones (c) Concrete (d) Timber
6. Match the name of stone with its application

Name of Stone	Application
Granite	Flooring
Lime Stone	Ornamental Work
Marble	Sea Walls
Slate	Cement Manufacturing

7. Which material keeps the structure cool in summers and warm in winters?
(a) Timber (b) Bricks (c) Bamboo (d) Clay
8. Which of the following is most eco-friendly construction material?
(a) Steel (b) Concrete (c) Mud (d) Glass
9. Concrete is strong in
(a) Compression (b) Tensile (c) Shear (d) None of the above
10. Bitumen is a by-product of which industry
(a) Paper (b) Chemical (c) Fertilizer (d) Distillery

11. Which of the following is an organic building material?
(a) Lime (b) Wood (c) Gypsum (d) Mud

12. Which material is used for external plastering works?
(a) Lime Plaster (b) Plaster of Paris (c) Mortar (d) None of the above

13. Which material out of the following is used for form work and shuttering?
(a) Concrete (b) Bricks (c) Stones (d) Timber

14. Which of the following is an artificial building material?
(a) Wood (b) Rocks (c) Sand (d) Concrete

15. Polypropylene fibres belong to which category of construction material
(a) Natural (b) Special (c) Finishing (d) Recycled

16. Which of the following is used to make partition walls?
(a) Bricks (b) Stones (c) Glass (d) Clay

17. Emulsion paints are used to provide
(a) Smooth Finish (b) Decorative Finish
(c) Water Proof Coating (d) All of the above

18. The by-product of thermal power plants is
(a) Fly Ash (b) Bagasse (c) Sludge (d) Metakaolin

19. POP is used for finishing of
(a) External Surface (b) Internal Surface
(c) Vertical Surface (d) All of the above

20. Steel is strong in
(a) Compression (b) Tension (c) Both (d) None of the above

ANSWERS																	
1.	(a)	2.	(b)	3.	(c)	4.	(a)	5.	(b)	6.	(a – 3, b – 4, c – 2, d – 1)					7.	(d)
8.	(c)	9.	(a)	10.	(d)	11.	(b)	12.	(a)	13.	(d)	14.	(d)	15.	(b)	16.	(c)
17.	(d)	18.	(a)	19.	(b)	20.	(c)										

Short Answer Type Questions

1. Discuss the scope of construction materials.
2. List of construction materials used in transportation engineering.
3. What are various types of construction material?
4. State the desired properties for a construction material.
5. Define serviceability.
6. Give some examples of natural construction materials.

7. State some applications of construction material in irrigation engineering.
8. Name any four artificial construction materials.
9. What are various types of strength considered for selection of a building material?
10. Write a short note on flyash.

Long Answer Type Questions

11. Explain the use of construction material in varied sectors of Civil Engineering.
12. Classify construction materialson the basis of their existence.
13. What are the factors considered for selection of construction materials? Explain.
14. Why is selection of appropriate construction material important for a construction project?
15. Write an explanatory note on finishing construction materials.

PRACTICAL

Aim: Identify the available construction materials in the laboratory on the basis of their sources.

Materials Required: Sample of various construction materials.

Theory: The construction materials can be identified and classified on the basis of their source of availability. The data for some of the common construction materials is as follows:

Material	Source	Category
Brick	Made in Kiln	Artificial
Clay	Natural Occurring in Soil	Natural
Mud	Natural Occurring in Soil	Natural
Timber	Exacted from Trees	Natural
Steel	Furnace	Artificial
Stone	Blasting of Rocks	Natural
Glass	Made in factories	Artificial
Concrete	Mixture of Cement, Sand, Aggregate and Water	Artificial
POP	Heating of Gypsum	Finishing
Lime Plaster	Mixture of Lime and Sand	Finishing

Procedure:

1. Collect the sample of available construction materials.
2. Arrange the construction materials on table.
3. Observe the construction materials closely to know their source.
4. Record the observations carefully.

Observations:

Sr. No.	Material Name	Source
01.	M1	
02.	M2	
03.	M3	
04.	M4	
05.	M5	

Result: The given material belongs to following categories.

Sr. No.	Material Name	Category (Natural/Artificial/Recycled/Special/ Finishing)
01.	M1	
02.	M2	
03.	M3	
04.	M4	
05.	M5	

Conclusion: The given construction materials can be identified and classified on the basis of their source of availability.

Precautious:

1. The given material should be in original shape and size.
2. Any additional covering on the material sample should be removed before observing it.
3. Minimum of three samples of a material should be observed.

KNOW MORE

The use of construction materials is as old as time immemorial. Building materials are the most important component of an infrastructure project as it involves huge cost and affects the life of a project directly. History is available for building construction materials for some thousands of years. Housing and shelter have always been on priority of human demands. Initially, temporary structures made up of leaves, tree stem, branches, straw, bushes were used as houses. Later came the era of stones, clay, mud and timber. The biggest advantage of these products was that these materials were available in abundance in nature. Mud and clay were popular because of their ease of moulding and could be held together using straw, sticks and other natural products like cattle dung. Wood was other common construction material because of its strength and ease to handle. Later there was a demand for permanent structures. Civilization shifted from tents and huts to permanent houses made up of cement, concrete, steel and bricks. The invent of these construction materials brought a revolution in construction industry as these materials were superior in terms of their strength, durability and serviceability. The use of these artificial construction materials are energy extensive and involves huge construction cost. Hence, the selection of proper building material has great significance. The growth of a country is developed by the kind of infrastructure it has. The modern construction has shifted from single-one-storey houses to skyscrapers and high rise buildings. These projects involve use of many more special construction materials like glass, steel, metals, PVC, PPFs etc. As the rising index of construction activities is putting a lot of stress on environment by depletion of natural resources on one side and release of harmful pollutants and waste generation on other side, there is a need to look for alternative construction materials. The research and trial activities on use of recycled products as construction materials can be a successful solution to solve the present problems of construction materials. Many successful examples of the same are available in the history.

REFERENCES AND SUGGESTED READINGS

Varghese, P.C., Building Materials, PHI learning, New Delhi.

Rangwala, S.C., Engineering Materials, Charator Publisher, Ahemdabad.

SOME INQUISITIVE EXAMPLES

1. **English Channel Tunnel (1994):** The English Channel Tunnel connects the shore of Kent in UK with Pas-De-Calais in France. It is designed to carry high speed passenger train, road vehicles and international goods trains at a depth of 115m below sea level at its deepest point. It is one of the highest value engineering projects.
2. **Brooklyn Bridge (1883):** It is one of the oldest bridges in United States. It is steel wire suspension bridge build with a cost of $15.5 million. It is a historic icon for New York City.
3. **Great Pyramid of Giza:** It is one of the seven wonders of ancient world. It was tallest structure in the world for over 3800 years. It is estimated that 5.5 million tonnes of lime stone, 5 lacs tonnes of mortar and 8 thousand tonnes of granite costing $5 billion were used in its construction.
4. **Bandra Worli Sea link, Mumbai:** It is an eight lane cable stayed bridge constructed in Arabian Sea to join two suburbs of Mumbai. It is an engineering marvel and an architectural wonder.
5. **Burj Khalifa, Dubai (2010):** The tallest structure in the world, Burj Kalifa is the most iconic building. Its construction started in 2004 and was completed in 2010. It is 828 m tall with 163 floors and having highest outdoor observatory deck. An estimated amount of 3 lacs cubic meter of concrete and 39 thousand tonnes of steel were used in its construction. The total aluminium used on Burj Kalifa is equivalent to five A380 aircrafts. It involves a world record of installation of glass façade at the height of 512 meters. It includes more than 380 skilled civil engineers in its construction.

2 Natural Construction Materials

UNIT SPECIFICS

This unit discusses the following aspects:

- Knowing various natural construction materials.
- Understanding the use of specific natural construction materials in construction.
- Detailed classification of various natural construction materials.
- Knowing various methods adopted to work with specific natural materials.
- Understanding the use of tools for natural construction materials.

Various experimentation procedures on natural construction materials are also explained in the unit to enhance the knowledge of students on the use of construction materials in an appropriate form. Some examples from history have been quoted in this chapter to create an inquisitive thought amongst the users about available natural construction materials. The additional list of suggested readings and references is given at the end of the unit to satisfy curiosity on the topic. This chapter imparts the knowledge on characteristics and use of natural construction materials like stones, timber, bituminous materials, lime, soil, sand and aggregates.

At the end of the unit, several multiple choice questions, short answer type questions and long answer type questions have been stated to assess the knowledge gained by the users. The framework blooms taxonomy i.e., remember, understand, apply, analysis and create have been kept in mind while making the questions. Assignment and quizzes can be framed using these questions. Further, many QR codes, provided throughout the unit, can be scanned to get deeper knowledge on the topic and to create interest on the subject. The section named 'Know more' acts as an icing on the cake and provides supplementary information on the characteristics and uses of natural construction materials. The unit summary at the end of the chapter gives a quick revision on 'what we have learnt' and generates creative and inquisitive nature among the students.

RATIONALE

This unit on natural construction materials helps students understand which materials are available naturally in our surroundings that could be used in the construction. It further imparts knowledge on the use of every natural construction material for the specific purposes. Besides, the characteristics desired for good natural construction material have been stated in the unit to help user select an appropriate material for a specific purpose at the site for a project. General properties, types and use of various natural construction materials give a comprehensive knowledge to the students on the topic. The unit further specifies the advantages and limitations on the use of various natural construction materials. Understanding the benefits available and the challenges faced in use of each natural construction material, ensures effective and efficient selection of construction materials. The use of natural construction materials has been popular since pre-history times but, at present, there are popular arguments on suppressing the use of natural construction materials and looking for alternative materials for construction. The advantages of natural construction materials make construction industry inclined towards their usage and at the same time their limitations provoke the industry to look for alternatives. A rationale on the use of natural construction materials can be sought by civil engineers after understanding this chapter thoroughly.

PRE-REQUISITE

Basic knowledge on construction materials (i.e., Unit-I)

UNIT OUTCOMES

Following is the list of outcomes from unit on Natural Construction Materials:

U2-O1: Define and understand various natural construction materials.

U2-O2: Understand the desired properties of natural construction materials.

U2-O3: Understand the types of various natural construction materials.

U2-O4: Know the uses of each natural construction materials.

U2-O5: Apply the knowledge gained to select appropriate natural construction material for a specific purpose.

Unit-2 Outcomes	EXPECTED MAPPING WITH COURSE OUTCOMES (1- Weak Correlation; 2- Medium Correlation; 3- Strong Correlation)				
	CO-1	CO-2	CO-3	CO-4	CO-5
U2-O1	3	3	1	2	2
U2-O2	2	3	1	1	1
U2-O3	2	3	1	2	1
U2-O4	3	3	1	2	1
U2-O5	3	3	1	3	1

2.1 BUILDING STONES

Stones are considered as one of the popular building material due to their easy availability in nature from rocks. The earth's crust is made up of rocks. Rocks are nothing but mineral mass, having more or less homogeneous uniform composition. Natural rocks are the main source for getting building stones and also for producing mineral binding materials like lime, gypsum and cement by certain processes. Building stones have the advantage of good strength and durability. In older days construction, many examples of structures like temples, dams, retaining walls, abutments, monuments etc. are made up of stones.

2.1.1 Good Building Stone

Stones are used in varied construction projects. Different civil engineering projects require several types of stone. Hence, it is important to check the suitability of a particular type of stone for a specific project which is having different loading and environmental conditions. However, some of the typical characteristics on which the quality of stone depends are discussed as below:

1. **Appearance:** Building stones should have pleasing appearance and should be of uniform colour. This quality becomes more important where stones are to be used for decorative works and placed for aesthetic beauty of the structure. Light coloured stones are superior to dark coloured stones in terms of their resistance to weathering action.
2. **Strength:** Most types of stones have reasonably good compressive strength. Due to this reason, stones are mostly used as a compression member. Igneous rocks have a better strength as compared to sedimentary rocks. Metamorphic rocks also possess relatively good strength.
3. **Weight:** Stones should have less porosity and more compactness. This property gives high weight to stone and hence, stones have high compressive strength.
4. **Hardness and Toughness:** Hardness is resistance to scratching and toughness is resistance to breaking. Stones with good hardness and toughness have resistance to wear and tear, specially required for road projects.
5. **Porosity and absorption:** All stones have some porosity. However, porosity, in stones, is not a desirable property as absorption of rainwater by stones leads to their deterioration. Also, in cold climates, the freezing of absorbed water in stones may cause cracks in stones.
6. **Compactness:** Building stones should have good compactness as this property helps a stone to withstand the effects of external agencies effectively.
7. **Grain Fineness:** Stones with fine grained structure are easy to carve/dress and hence, are suitable for moulding. However, non-crystalline stones tend to disintegrate/break under the action of external natural agencies.
8. **Fire Resistance:** Stones must be fire resistant. Fire resistance means that the stone should be able to retain its shape and size on exposure to fire. Most of the stones can resist fire up to very high temperature; say of the order of 800°C.
9. **Durability:** Durability of stone refers to its longevity of life. Stones which have acid resistance and negligible water absorption are durable. Stones having compact and homogenous nature have better durability.
10. **Dressing:** The process of shaping of stones for some decorative purpose is called dressing. Soft stones like slates and limestone are easy to dress. Hard stones cannot be dressed easily.

2.1.2 Requirements of Good Building Stone

(i) A good stone should be free from soft patches, cracks, cavities and is of uniform colour and texture. This ensures good appearance of stone.

(ii) The building stone should be compact to ensure good strength and high density.

(iii) A good building stone should be hard, strong and durable. It should have good weather resistance, acidic resistance, fire resistance and negligible water absorption.

(iv) The stone should be easily and economically available in bulk. Most of the construction projects require huge quantities of stones and hence, the stones should be available near the project otherwise the cost of transportation of stones is going to be very high and can make the project uneconomical.

(v) The good stones should be easily workable so that it can be dressed/ moulded into desired shape.

The following table gives the value for some of the desirable properties of good building stone:

Table 2.1: Desirable limits for Properties of Good Building Stone

Property	Desirable Limit
Appearance	Uniform Colour
Strength	> 100N/mm^2
Specific Gravity	> 2.70
Hardness i.e. wearing Resistance	< 3%
Water Absorption	> 0.6% by weight
Compactness	Fine crystalline structure
Fire Resistance	> 800°C

2.1.3 Quarrying of Building Stones

The process of taking out stones from rock beds is called quarrying and the places from where stone are extracted are known quarry.

2.1.3.1 Methods and Tools of Quarrying

(i) **Digging:** Digging is the simplest method to obtain stones and is applicable for soft rocks. Digging of soft rocks is carried out with the help of hand tools like chisels, hammers etc.

(ii) **Heating:** The exposed part of some rocks is separated from the main mass by heating process which is carried out by burning some fuels on the exposed surface of the rocks. Due to continuous heating for hours, there is an unequal expansion of the rock, and that part of the rock gets separated and stones are obtained from it.

(iii) **Wedging:** There are some rocks which are soft and stratified i.e., these are formed in layers which can easily be split. Such rocks can be drawn to stones by wedging. The tools used for wedging are pickaxes and crow basis. Wedging is a simple method which is successfully used for quarrying of limestone, marbles and slates.

(iv) **Blasting:** Blasting is the most typical method of quarrying. It is used for hard and compact rocks. Blasting is carried out in various stages. Firstly, holes are drilled in rocks with the help of steel bars with sharp ends or by using drilling machine. Then, the drilled holes are filled carefully with suitable explosives. These explosives are then tempted to avoid blasting of explosives within the holes before firing. The fuse for firing is placed carefully and later fired. When the explosion takes place, the masses of stones all around the hole are removed and obtained.

2.1.4 Dressing of Stones

The stones obtained from quarrying have a very rough surface and irregular shape. The process of finishing the stones to obtain a regular shape and smooth surface is called dressing of stones. Hard stones are difficult to dress. The benefits of dressing are:

(a) Dressing results in good appearance and regular shape of stones.

(b) Dressed stones are easy to handle in terms of their transportation and usage.

(c) Dressing ensures proper bedding in stone masonry and suit building requirements.

2.1.4.1 Methods of Dressing of Stone

(i) **Quarry Dressing:** When the stone dressing operations are carried out at the quarry site itself, it is called quarry dressing. This helps in reducing the cost of transportation.

(ii) **Site Dressing:** When stones are shaped and surface finished at the site of usage, it is called site dressing. This method of dressing is preferred for more effective dressing of stones which ensures desired suitable size and surface finish.

2.1.5 Tools used for Dressing of Stones

Dressing of stones involves use of many tools as shown in Fig. 2.1. Now a days, mechanical tools are preferred over manual one to save time for dressing of stones. Some of the tools used include:

- **Hammer:** Hammer is the most commonly used tool for dressing of stones. Hammered dressed stones do not have any shape edges or irregular corners and hence, are well suited for masonry works. Some of the common types of hammers are Club Hammer, Spall Hammer, Mash Hammer and Wallers Hammer.
- **Chisel:** Chisels are sharp tools used to remove extra part of stone near their central mass. Chisel dressed stones are used in plinths and corners of the building by dressing in required shape and size.
- **Gad:** Gad is a metallic tool made up of iron, used to split the stones in pieces.
- **Saw:** Various types of saws are used to dress the stones by cutting process. Hand saw is used to cut soft stones. Cross-cut saw is used to cut hard stones.
- **Pitching Tool:** Pitching tool is used to get stones of required size by removing layers of masses from original stone.
- **Square Set:** It is a tool used to set edges at right angles for stones.
- **Punching and Pointing Tool:** Punch tool is used for rough dressing of soft stone whereas pointing tool is used to make holes or rough dressing in hard stones.
- **Drafting Chisel:** These are special purpose chisel used to dress the stones very finely. Drafting of stones refers to maintaining fine edges, cuts and shapes in stones.

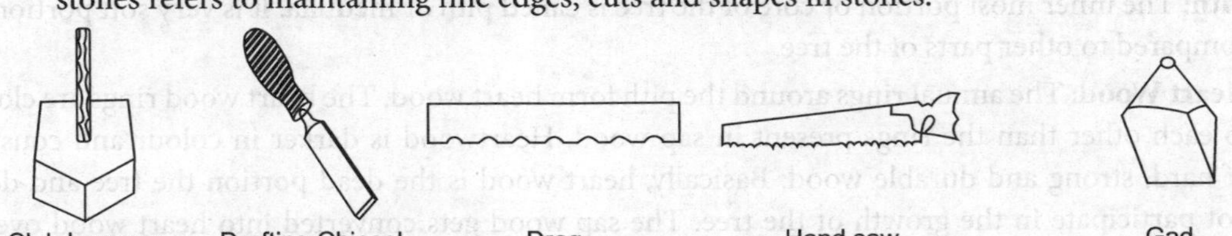

Club hammer | Drafting Chiesel | Drag | Hand saw | Gad

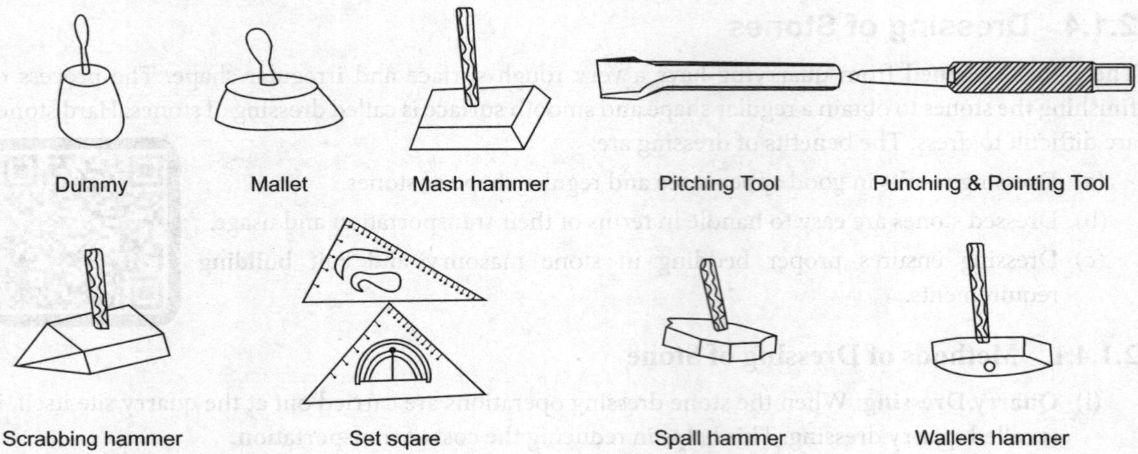

Figure 2.1: Tools for Dressing of Stones

2.2 TIMBER

The wood which is suitable for construction and other engineering purposes is called Timber. In building construction, timber is used in doors, windows, partitions, shelves, furniture and railing etc. Generally, timber is obtained from two types of trees:

- **Endogenous Trees:** The trees which grow inwards e.g., bamboo and palms. These are found in tropics and have limited use in engineering.
- **Exogenous Trees:** The trees which grow outwards i.e., there stems get thicken with growth e.g. Teak, Babul, Sal etc. These are mainly used for engineering and building construction purpose.

Timber is an expensive material and should be carefully selected for a specific purpose. Timber is most commonly used in the form of plywood or raw wood. Ply board is obtained by pressing together several layers of wooden sheets and ply board refers to board made up of agro waste, cellulose etc, blended with adhesives to give a solid board. Most popular woods for doors and windows include Teak, Sal and Deodar. Teak wood is considered superior over other wood for the advantage of its natural look, durability and long lasting characteristics.

2.2.1 Structure of Timber

The structure of timber which is visible to naked eye is called its macro-structure. The cross-section of timber has ring like structures, called annular rings which can be counted to know the age of the tree. The structure of timber (Fig. 2.2) shows following parts:

1. **Pith:** The inner most portion or core of the tree is called pith or medulla. It is very soft portion as compared to other parts of the tree.
2. **Heart Wood:** The annual rings around the pith form heart wood. The heart wood rings are closer to each other than the rings present in sap wood. Heartwood is darker in colour and consists of hard, strong and durable wood. Basically, heart wood is the dead portion the tree and does not participate in the growth of the tree. The sap wood gets converted into heart wood over a

period of time of the order of 15 – 20 years. Heart wood is the best portion of a tree to be used for engineering purpose.

3. **Sap Wood:** The outer annual sings around the heart wood are known as sap wood. This portion is generally lighter in colour and weight than heart wood. Sap wood has a lot of moisture content and it participates actively in the growth of the tree. Sap wood is usually not used for construction purposes as it is liable to decay fast.
4. **Cambium Layer:** The thin layer of sap i.e., plant juice and moisture, present between sap wood and inner bark is known as cambium layer. This layer is nothing but immature sap wood.
5. **Inner Bark:** The inner bark is a skinny layer around cambium layer which protects cambium layer from any injury and ensures supply of sap juice to other parts of the tree.
6. **Outer Bark:** The outer most protective layer or skin of a tree is called outer bark. Outer bark protects the tree from high temperature and any kind of mechanical harm.
7. **Medullar Rays:** The thin radial fibres which extend from pith to cambium layer are called medullary rays. The function of medullary rays is to hold the annual rings of sap wood and heart wood together and to supply the stored food to the parts of the tree according to the necessity.

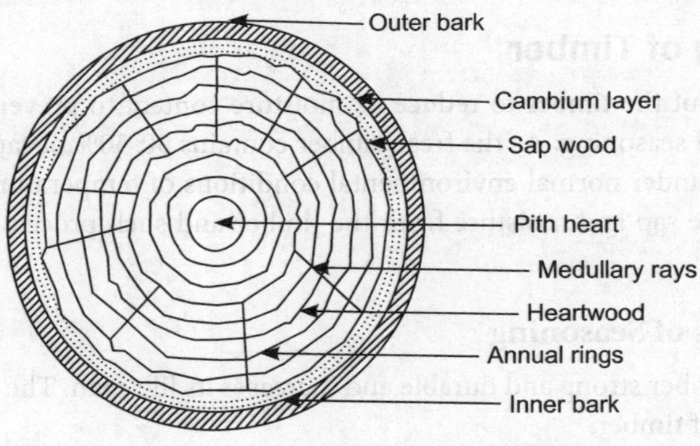

Figure: 2.2: Structure of Timber

2.2.2 General Properties of Good Timber

A good timber should be strong, tough and durable. A good timber should possess following characteristics:

(i) **Appearance:** A good timber should have uniform and compact appearance. It should have straight fibres which are considered pleasing to sight and should have shiny appearance.

(ii) **Colour:** The timber is considered good, if it is dark in colour. Dark colour is taken as a symbol of good strength.

(iii) **Defect-Free:** A good timber should be free from defects like knots, flaws, shakes etc. Also, a timber should have sweet smell and it should give good sound when struck to ensure that it is defect free.

(iv) **Strength:** A good timber should be strong enough to bear the loads applied to it. It should be strong against compressive loads as well as against loads applied in transverse direction. Timber with good strength is used as structural members like beam. Rafter, joist etc.

(v) **Durability:** A good timber should be durable against the attack by fungi, insects, chemicals etc. and should also be able to resist the action of physical and mechanical agencies.

(vi) **Weather Resistance:** A good timber should be capable of offering good resistance against weathering forces like alternate wetting and drying, alternate freezing and thawing (cooling and Heating) conditions, temperature variations, wind effects etc.

(vii) **Fire Resistance:** Most of the timber are generally not fire resistance. However, a good timber should have sufficient fire resistance i.e., it should not catch fire easily. A dense compact wood has better fire resistance.

(viii) **Elasticity:** Elasticity refers to the property of regaining original shape and size when loads are removed. A good timber should be capable of regaining its original shape without any permanent deformation when loads are removed.

(ix) **Workable:** A good timber should be easy to work with tools like saw, planning tools etc. and it should not clog the teeth of the saw.

(x) **Toughness and Abrasion:** A good timber should be able to resist shocks due to vibration. Also, it should not deteriorate due to mechanical wear i.e., due to rubbing action of the loads.

2.2.3 Seasoning of Timber

The process of drying out the timber to reduce its moisture content to prevent it from possible decay or fermentation is called seasoning. As the fresh timber contains 30-50% of sap and water, it is liable to decay or bending even under normal environmental conditions of temperature variations. Hence, it is important to remove the sap and moisture from the timber and such processed are collectively called seasoning of timber.

2.2.3.1 Advantages of Seasoning

Seasoning makes the timber strong and durable and increases its life span. The following advantages are achieved by seasoning of timber:

(i) Seasoning decreases the volume and weight of the timber and hence makes it easy to handle and transport.

(ii) Seasoning of timber improves the strength, hardness and toughness of the timber.

(iii) Seasoning results in reduced moisture content in timber and hence improves the resistance to decay due to attack by insects and fungi.

(iv) Seasoned timber is more workable as it has reduced tendency to bend, wrap or crack.

(v) Seasoning of timber renders it fit for decorative works like polishing, painting and varnishing etc.

2.2.3.2 Methods of Seasoning

The methods of seasoning of timber are broadly classified into two categories:

(i) **Natural Seasoning or Air Seasoning:** Natural seasoning of timber is a simple, cheap and least care demanding process. In this method timber, in the form of logs or planks or post, is dried by direct action of air, water or sun. The timber logs are stacked in cross wise direction in alternate layers to ensure free circulation of air around them. This type of seasoning is very slow and takes few months to an year depending upon the kind of wood and its thickness. As there is no control on temperature and humidity in natural seasoning, the desired quality of seasoning is not attained.

(ii) **Artificial Seasoning:** Artificial seasoning, besides being a fast process, has the advantages of controlled temperature and humidity and hence, the timber can be dried under control conditions to attain desired moisture content level. The various methods used for artificial seasoning are:

(a) **Kiln Seasoning:** In this method, timber is stacked in a chamber in a kiln and hot air is freely circulated around it for 3 – 6 days. In this way, the timber is dried to desired moisture content level at a much faster rate under controlled conditions of temperature and humidity. Kiln seasoning is an expensive method requiring skilled supervisor, but it gives a well seasoned uniform timber.

(b) **Chemical Seasoning:** In this method, the timber is immersed in the solution of suitable salts like ammonium carbonate or urea which absorbs the moisture content from inside of the timber and then timber is left for air seasoning for exterior surface. This method takes 30 – 40 days.

(c) **Electric Seasoning:** In this method, high frequency currents are passed through the timber which helps in drying of sap and moisture. This method is the fastest method of seasoning and takes 5 – 8 hours only, however, it is a very costly method and not suited for large scale seasoning.

(d) **Water Seasoning:** In this method, logs of freshly felled trees are fully immersed in running stream of water for about 3 – 4 weeks. The thicker end of the log is aligned in upstream direction and hence, the sap of the timber is washed away by water. Then the logs are taken out of the water and kept in stacked form for drying out by free circulation of air around it. This method is generally not preferred as continuous immersion of timber in water makes it weak and brittle and hence reduces the durability of timber.

2.2.4 Defects in Timber

Timbers of all types have a chance to face some natural defects depending upon the environmental conditions to which the trees are exposed. Hence defects should be avoided while converting wood into useful timber. Some of the common defects in timber (Fig. 2.3) are:

(i) **Heart Shakes:** Heart shakes are cracks or splits starting from the central part of timber i.e., pith and extending along medullary rays up to sap wood. Heart shakes are caused due to shrinkage drying of inner part of timber i.e., heart wood due to age. A heart shake running straight across the trunk is not considered serious.

(ii) **Star Shake:** The cracks or splits that run from outer bark to sap wood are known as star shakes. These cracks are caused due to scorching heat or frost action on the tree. Star shakes are more common at outer ends and reduce towards centre.

(iii) **Cup Shakes:** Cup shakes are curved splits along the annular rings (partly or wholly) which are caused to rupture of tissues. Whenever trees face unequal growth due to sudden expansion or contraction due to winds, cup shakes are prominent.

(iv) **Radial Shakes:** Radial shakes look similar to star shakes but are caused when felled tree is exposed to sun during seasoning. These cracks are generally, fine, irregular and numerous. Radial shakes originate from outer surface and extends towards entre.

(v) **Rind Galls:** The curved enlarged swelling found in trees when its branches are improperly cut off are called ring galls. The wood is hardened at the formation of rind galls.

(vi) **Wind Cracks:** Wind cracks are splits on the outer bark of the timber due to shrinkage when the exterior of the timber is exposed to atmospheric conditions of sun and wind.

(vii) **Knots:** Knots are the bases or roots of small branches of trees. Knots are not harmful. Knots break the continuity of the fibres and hence are not appealing to the sight. Timber with large dead knots should be avoided for the reasons of poor appearance and strength.

(viii) **Dead Wood:** Timber which is obtained from a tree felled after maturity, is called dead wood. It is light in weight and has less strength and durability. It is slightly reddish in colour.

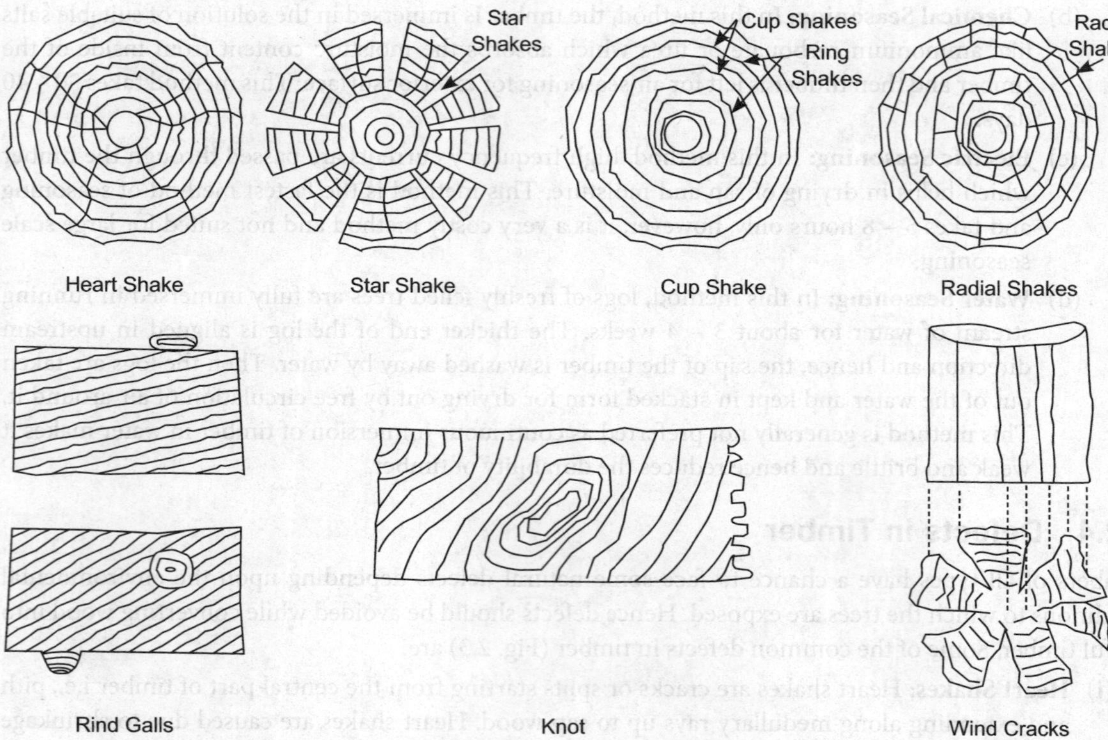

Figure 2.3: Defects in Timber

2.2.5 Bamboo in Construction

Bamboo is one of the sustainable building materials obtained from endogenous tree. The advantages that make bamboo suitable for construction are:

(i) Bamboo is a very strong fibrous timber.

(ii) The compressive strength of bamboo is almost twice that of concrete.

(iii) The tensile strength of bamboo is as good as that of steel.

(iv) Bamboo fibre has best shear strength than any other wood.

(v) Bamboo is easy to bend and curve without breaking which makes it easy workable.

However, even with so many advantages, bamboo is not a popular building material. It has its own limitations too. These include- bamboo is vulnerable to attack by termites and fungi which lowers the durability and life span of the structures. Bamboo can be treated with borax boric acid solution to preserve it from the attacks and increase its life span. But use of chemicals is not promoted due to their harmful effects on environment. Hence, bamboo is generally used in construction for non-structural members and furniture only.

2.3 ASPHALT, BITUMEN AND TAR

2.3.1 Asphalt

Asphalt is defined as a mixture of bitumen and inert materials like sand, gravel and crushed stone. Bitumen in asphalt acts as a binding material and the other materials act as filler. Asphalt is blackish brown in colour. It is in solid state as low temperature and at high temperature of the order of 50°C or above, it is in liquid state. Asphalt is obtained in two forms:

(a) **Natural Asphalt:** Natural asphalt is found in nature in some lakes and rocks formation. Lake asphalt consists of 40 – 70% pure bitumen and is used road laying after boiling and removal of impurities. Rock asphalt has only 10 – 15% of pure bitumen and it is used for paving tiles etc.

(b) **Residual Asphalt:** It is produced by fractional distillation of crude petroleum.

2.3.1.1 Forms of Asphalt

(i) **Cutback Asphalt:** It is obtained by dissolving asphalt in a volatile solvent. Cutback asphalt is in liquid state and is used for repairing roofs and floors in cold condition. It is also used for preparing bituminous paints.

(ii) **Asphalt Emulsion:** Asphalt emulsion in obtained by mixing asphalt in 50 – 60% water in presence of 1% emulsifying agent. When used, the water from emulsion evaporates and results a waterproof film. This form is used for damp-proofing purposes in cold regions.

(iii) **Asphalt Cement:** It is obtained by blowing air at high temperature through natural asphalt. The asphalt, on oxidizing, results in a plastic product which has high resistance to atmospheric agencies. It is used for roofing, flooring, water proofing and as a filler material for expansion joints.

(iv) **Mastic Asphalt**: It is obtained by heating natural asphalt with sand and mineral fillers. The resulting product is an impermeable solid or semi solid matter which can be applied in construction of roofs, floor, paving, damp proof course etc.

2.3.2 Bitumen

Bitumen is produced by fractional distillation of crude petroleum. The composition of bitumen is 87% carbon, 11% hydrogen and 2% oxygen. Bitumen is present in asphalt and tar also. The various forms of bitumen are:

(i) **Cutback Bitumen:** Cutback bitumen is obtained by fluxing of asphaltic bitumen in presence of coal tar or petroleum. It is used for manufacturing paints for cold regions.

(ii) **Bitumen Emulsion:** It is obtained by mixing very finely divided bitumen in an aqueous medium in presence of stabilizing agents.

(iii) **Plastic Bitumen:** It is obtained by mixing of bitumen, thinner and inert fillers like asbestos fibres etc. The filler material is generally kept between 40-50% of bitumen. Plastic bitumen is suitable as a sealing material for leakages and filling cracks in masonry structures.

(iv) **Blown Bitumen:** It is obtained by passing air under high temperature and pressure through bitumen. It is available in ten grades. Blown bitumen is used as a damp proofing layer, heat insulating material, joint fillers and leak sealant.

(v) **Straight Run Bitumen:** It is obtained by distillation of bitumen up to a definite viscosity. No further treatment is done on straight run bitumen. It is used as a water proofing compound.

2.3.3 Tar

Tar is a highly viscous liquid containing high amount of carbon content. It is used for forming a surface layer for pavements, wood preservation etc. The various forms of tar are –

(i) **Coal Tar:** It is a residue obtained during production of coal gas. It is a dense black colour strong smelling liquid which is used in preserving timber and laying macadam roads etc.

(ii) **Wood Tar:** The distillation of resinous woods like pine provides a sort of oil. It is called wood tar. It is used as a preservative for timber.

(iii) **Mineral Tar:** Mineral tar is obtained from Kerogens. Kerogens are present in bituminous shales i.e., rocks. Mineral tar has very less volatile content.

2.4 LIME

Lime is a basic building material used in many forms in construction activities. Lime is basically a product of calcium. It is obtained by burning of stones, rich in calcium carbonate, at very high temperature of order of 900°C or more for several hours. It is produced in powdery form and used in hydrated or non-hydrated forms.

2.4.1 Properties of Lime

The important properties of limes which make it suitable for construction purposes are:

(i) **Cementing Capability:** The lime (Calcium Oxide) in presence of moisture reacts with carbon dioxide (CO_2) present in atmosphere and results in formation of calcium carbonate ($CaCO_3$) which has cementing properties. This process is called carbonation. This property helps in using lime as a mortar for brick masonry construction.

(ii) **Acid Resistance:** Lime is alkaline in natural and hence, it has higher acid resistance.

(iii) **Pozzolanic Property:** Carbonation of lime results in cementitious product. Lime has good pozzolanic properties which make it fit for structural elements.

(iv) **Sealing Property:** When lime mortar mix is passed by gaseous carbon dioxide, it results in precipitates of calcium carbonates which seal the micro-cracks in building. The other properties of lime like fineness, setting time and soundness are comparable with cement but the compressive strength of lime is less (say 2 – 3N/m^2) than that of cement.

2.4.2 Types of Lime

Different types of limes are used for construction purposes. The various types of lime mainly differ from each other on the basis of process of hardening. Some of the common types of lime used in construction are:

(i) **Quick Lime:** It is also called caustic lime. Quick lime is calcium oxide (CaO) obtained by burning and calcinations of pure lime stone. It is the cheapest form of lime. It has a great affinity of moisture. Quick lime is used as a main ingredient in the manufacturing of cement. It is also used in treatment of drinking water.

(ii) **Slaked Lime:** Quick lime combined with water is called Slaked lime or hydrated lime. It is formed as slurry which is highly suitable for mortar applications. It is also used for plastering purpose and as a binder in cement production.

(iii) **Fat Lime:** The lime containing high content of calcium oxide is called fat lime. It is also known as rich lime or white lime or pure lime. When water is added to fat lime, its volume increases by two and a half times greater than that of quick lime. Fat lime is used in foundation or masonry walls to thicken it.

(iv) **Hydraulic Lime:** Hydraulic lime contains 30% silica, 5% alumina and some fractions of iron oxide. Its properties are similar to that of cement. It is also known as water lime. Hydraulic lime sets when water is added to it. It is used in mortar.

2.4.3 Uses of Lime in Construction

(i) Lime is used as a basic ingredient in manufacture of Ordinarily Portland Cement (OPC).

(ii) Lime is used in manufacture of steel to remove impurities.

(iii) Lime slurry is used as mortar in masonry works and plastering works.

(iv) Lime is used in large quantity for soil stabilization during construction of roads, pavements and building foundations.

(v) Lime is used in treatment of drinking water.

2.5 SOIL

Soil has been a popular building material since early ages of civilization. History has evidence of using mud as a construction material since Neolithic times. Soil in the form of mud is available in abundance almost everywhere. Soil or mud construction is preferred in relatively dry places. Soil is a natural combination of minerals with or without organic matter that can be separated by mechanical means easily. The soil for construction is the layer of earth collected after a depth of top 60 cm only. The top layer of ground is full of organic matter and it is unfit for use as a construction material. Below the top layer, there is a combination of sand and clay which can be used as a building material.

2.5.1 Types of Soil

The following are various types of soil and their suitability in construction.

(i) **Gravel:** Gravel are cohesionless small pieces of stone present in earth layer. Gravel can be suitably used in construction as aggregates. Their size various from the size of a pea (2 mm) to the size of an egg (60 mm). Such soil does not have the property of swelling or shrinkage on wetting and drying and hence, it provides a very good foundation surface for a building.

(ii) **Sand:** Sand is very much similar to gravel but the particle size is much smaller. Sand particles are cohesionless aggregates with size ranging from 0.06 mm to 2 mm, but each particle is visible to the eye. These soils also do not process the property of swelling and shrinkage on wetting and drying. Sand particles are permeable and do not allow water level to rise through the capillary action. Sandy soil provides the best foundation surface for the building.

(iii) **Silt:** Silt is very similar to sand, but the particle sizes are much smaller and not visible as individual particles or grains. The size of silt particles lies between 0.002 mm to 0.06 mm with no or very little plasticity. Silty soil is not considered as a good option for foundation surface as it shows some tendency of swelling and shrinkage on wetting and drying. Also, silt has some permeability, and many allow rise in water level in foundation.

(iv) **Clay:** Clay is composed of microscopic particles of size less than 0.002 mm, formed by chemical decomposition or disintegration of rocks. Clay soil become sticky when wet and is very hard in dry state. Clays have very low permeability and these exhibits the property of swelling on wetting and shrinkage on drying. Clay soil has very low load bearing capacity and hence not suitable for foundation of building. The structures constructed on clay soil are liable to undergo settlement.

(v) **Organic Soil:** The uppermost layer of soil is formed by decomposition of organic matter like fallen leaves, plants and vegetable matter. Such soil is called organic soil and is not fit for construction. The soil gives foul smell due to decaying matter and becomes spongy when wet. The soil is dark in colour and exhibits dampness.

2.6 SAND

Sand is one of the naturally occurring granular particles formed by finely divided rocks. These particles may be round, angular or flat in shape and their size is compulsorily less than 4.75 mm. The main constituent of sand is silica, a form of quartz. The main natural sources of sand are river, sea and pit. However, these days, artificial sand, obtained by fine crushing of granite or basalt rock is also used as an effective alternate to river sand.

2.6.1 Types of Sand

Based on grain size distribution, Sand is broadly classified as follows:

(i) **Fine Sand:** The sand having 90% of its particles size greater than 0.06 mm and less than 0.2 mm is called fine sand. This sand is best suited for plastering as it ensures a smooth surface.

(ii) **Medium Sand:** The Sand with 90% of its particles in size range of 0.2 mm to 0.6 mm is called medium sand. This sand is used in preparing mortar.

(iii) **Coarse Sand:** The sand with its 90% particles of size grater then 0.6 mm and less than 2 mm belongs to coarse sand. It is usually used in masonry works.

(iv) **Gravel Sand:** The sand is also called fine aggregate. It has size of particles between 2 mm to 4.75 mm. This sand is used for concrete preparation.

2.6.2 Properties of Good Sand

For sand to be used as a building material, and good sand should have the following properties:

(i) The sand should be chemically inert.

(ii) It should be clean from deleterious materials.

(iii) Sand should be free from organic matter.

(iv) The particles of good sand should be sharp, angular and durable.

(v) A good sand is well graded i.e., it should contain particles of all sizes in suitable proportions.

(vi) It should be free from coatings of clay and silt.

2.6.3 Uses of Sand

In construction works, sand is basically used for the following purpose:

(i) **Brick Masonry Works:** Fine/Medium sand with silt content less than 4% is used in preparing mortar for masonry work.

(ii) **Plastering Work:** Fine sand with silt content not more than 4% and fineness modulus less than 1.5, is used for plastering work.

(iii) **Concreting Works:** Coarse sand with fineness modulus between 2.5 to 3.5 is used as fine aggregate for concreting purposes.

2.7 COARSE AGGREGATES

Coarse aggregate is a general term used for chemically inert materials which are added to make concrete. These have a good bonding with cement-sand paste and provide strength to the concrete. Coarse aggregates are defined as particles with size greater than 4.75mm. These are, mostly, natural stone or gravel but can even be artificially processed aggregates like broken bricks or crushed air cooled blast furnace slag. Coarse aggregates can be classified on the basis of various parameters like size, shape, texture, origin, density etc.

2.7.1 Classification Based on Size

(i) **Single Size Aggregates:** If 85% to 100% particles of a lot of aggregates pass through a specified size of sieve, it is called single size aggregates.

(ii) **Graded Aggregate:** These aggregates contain proportion of different size of aggregates and hence give minimum voids and good strength in concrete.

(iii) **Gap-Graded Aggregate:** These are graded aggregates in which one particular size of aggregates is missing and hence it is called gap-graded.

2.7.2 Classification Based on Shape

(i) **Rounded:** The aggregates which have no edges and are round in shape due to attrition/rubbing effect of natural agencies or artificial means are called rounded aggregates.

(ii) **Irregular:** Naturally irregular or partially rounded aggregates by attrition are called irregular aggregates.

(iii) **Angular:** The aggregates having well defined edges are called angular aggregates.

(iv) **Flaky:** The aggregates whose thickness is relatively much smaller than their length and breadth are called flaky.

2.7.3 Classification Based on Texture

(i) **Smooth:** The aggregates having plain smooth glassy surface belong to this category.

(ii) **Granular:** The aggregates having grain surface are called granular aggregates.

(iii) **Crystalline:** The aggregates which possess crystal appearance on their surface.

(iv) **Honey Combed and Porous:** The aggregates which have pores on their surface and gives honeycomb appearance on their surface are called honey combed aggregates.

2.7.4 Classification Based on Origin

(i) **Natural:** Aggregates which are obtained from natural sources like riverbed, mines or quarries are called natural aggregates.

(ii) **Processed/Artificial Aggregates:** These aggregates are either obtained as a by-product or through processing. Common processed aggregates include burnt clay, industrial slag or broken bricks.

2.7.5 Classification Based on Density

(i) **Light Weight Aggregates:** These aggregates are generally porous in nature. The density of light weight aggregates is about 1000 – 1200kg/m^3. These types of aggregates are not used for structural loading carrying member of concrete. These are used to produce light weight concrete. These aggregates provide better thermal insulation and fire resistance.

(ii) **Standard Aggregates:** The normal aggregates which are available from riverbed or pit and are used in common structures works refers to standard aggregates. Their density is about 2400 – 2600kg/m^3. These aggregates generally contain varied sizes of particles and hence are graded ones. Standard aggregates are used in normal concrete works.

(iii) **High Density Aggregates:** Aggregates having density more than 2800kg/m^3 are called high density aggregates. There are used to form high strength concrete. The concrete made by high density aggregates is more dense, strong and durable. These aggregates are also used for very heavy weight concrete which is required to shield Gamma or X-ray radiations such as in nuclear reactors etc.

UNIT SUMMARY

- Natural Construction Materials
  - Stone
  - Timber
  - Asphalt and Bitumen
  - Lime
  - Soil
  - Sand
  - Aggregates
- Building Stone – Important Properties
  - Appearance
  - Strength
  - Weight
  - Hardness
  - Porosity
  - Compactness
  - Fire Resistance
  - Durability
- Quarrying Methods
  - Digging
  - Heating
  - Wedging
  - Blasting
- Dressing methods
  - Quarry Dressing
  - Site Dressing

- Timber Structure
  - Pith
  - Heart Wood
  - Sap Wood
  - Cambium Layer
  - Outer Bark
  - Annular Rings (Depicts Age of Tree)
  - Medullary Rays (Food Supply Chain for Tree)
- Properties of Good Timber
  - Appearance
  - Colour
  - Defect Free
  - Strength
  - Durability
  - Fire Resistance
  - Elasticity
  - Workable
  - Toughness
- Seasoning of Timber (Drying of Timber) Methods
  - Natural Seasoning (Air Seasoning)
  - Artificial Seasoning (Kiln Seasoning, Chemical Seasoning, Electric Seasoning, Water Seasoning)
- Defects in Timber
  - Heart Shakes (Cracks starting from centre)
  - Star Shakes (Cracks starting from outer bark)
  - Cup Shakes (Cracks curved along annular rings)
  - Rind Gals (Swelling in trees at cut off of branches)
  - Wind Cracks (Caused due to shrinkage by wind action)
  - Knots (Roots of branches – hamper the appearance)
  - Dead Wood (Felled tree timber with less strength and durability)
- Bamboo
  - Good Tensile Strength
  - Good Shear Strength
  - Easy to Bend and Curve
  - Easily Workable
- Asphalt, Bitumen, Tar
  - Asphalt: 40 – 50% pure bitumen, used in road construction
  - Bitumen: obtained from fractional distillation of crude petroleum. Used as damp proofing layer, Paints, Heat Insulation and Leak Sealant
  - Tar: Very high carbon content, used for preserving Timber and as surface layer in road construction
- Lime (CaO) Properties
  - Cementing Properties
  - Acid Resistance
  - Pozzolanic Property
  - Sealing Property
- Types of Lime
  - Quick Lime: Used in treatment of drinking water

- Slaked Lime: Used in plastering works and cement manufacturing
- Fat Lime: Used in foundations and masonry walls
- Hydraulic Lime: Used as mortar in masonry works

- Soil Types
  - Gravel: Size between 2 mm to 60 mm
  - Sand: Size between 0.06 mm to 2 mm
  - Silt: Size between 0.002 mm to 0.06 mm
  - Clay: Microscopic size less than 0.002 mm
  - Organic Soil: Upper most layer containing organic matter
- Sand Types
  - Fine Sand: Particle size between 0.06 mm to 0.2 mm
  - Medium Sand: Particle size between 0.2 mm to 0.6 mm
  - Coarse Sand: Particle size between 0.6 mm to 2 mm
- Uses of Sand
  - Brick masonry Mortar
  - Plastering
  - Concreting
- Coarse Aggregates – Classifications based on
  - Size: Single Graded, Well Graded, Gap Graded
  - Shape: Round, Irregular, Angular and Flaky
  - Texture: Smooth, Granular, Crystalline, Honey Combed
  - Origin: Natural, Artificial Aggregates
  - Density: Light Weight, Standard Weight, High Density Aggregates

EXERCISES

Multiple Choice Questions

1. A heavy stone is suitable for:
   (a) Arches (b) Rubble Masonry (c) Roads (d) Retaining Walls
2. A permissible value of water absorption for a good stone is
   (a) 5% (b) 10% (c) 15% (d) 20%
3. Chisel is a tool used for
   (a) Quarrying of Stones (b) Dressing of Stones
   (c) Testing of Stones (d) All the above
4. Process of obtaining plain edges with desired shape and size of stone is called
   (a) Quarrying (b) Seasoning (c) Dressing (d) Blasting
5. Crushing strength of a good building stone is not less than
   (a) 50 MPa (b) 100 MPa (c) 150 MPa (d) 200 MPa

6. Which tool is used to make holes in soft stones
(a) Pitching Tool (b) Punching Tool (c) Drafting Saw (d) Pointing Tool

7. Hard stones are cut using
(a) Gad (b) Hand Saw (c) Cross-Cut Saw (d) Chisel

8. Which of the following gives hard wood
(a) Neem (b) Deodar (c) Shisham (d) Chir

9. The core of the tree is called
(a) Pith (b) Heart Wood (c) Sap Wood (d) Cambium

10. Heart Shakes are caused due to
(a) Lack of ventilation (b) Age related shrinkage
(c) Alternate wetting and drying (d) Attack by environmental action

11. The moisture content for a well seasoned timber is
(a) 4 – 6% (b) 10 – 12% (d) 15 – 20% (d) 100%

12. The age of a tree can be known by examining
(a) Cambium Layer (b) Medullary Rays
(c) Annular Rings (d) Heart Wood

13. Radial Shakes are numerous cracks caused due to
(a) Lack of Ventilation (b) Exposure to Sun
(c) Attack by Termite (d) Alternate Freeze and Thaw Conditions

14. Bamboo has good
(a) Compressive Strength (b) Bonding Strength
(c) Tensile Strength (d) All of the above

15. Which of the following has good acidic resistance?
(a) Timber (b) Stone (c) Bitumen (d) Lime

16. Which is the cheapest form of lime?
(a) Quick Lime (b) Slaked Lime (c) Fat Lime (d) Hydraulic Lime

17. Which of the following is smallest in size?
(a) Sand (b) Silt (c) Clay (d) Gravel

18. Which sand is best suitable of plastering work?
(a) Fine Sand (b) Medium Sand
(c) Coarse Sand (d) Gravel Sand

19. Which of the following aggregates is best for concreting?
(a) Single Sized (b) Well Graded (c) Gap Graded (c) All of the above

20. Which of the following aggregates have low density?
(a) Glassy (b) Crystalline (c) Honey Combed (d) Flaking

ANSWERS																			
1.	(d)	2.	(a)	3.	(b)	4.	(c)	5.	(b)	6.	(d)	7.	(c)	8.	(c)	9.	(a)	10.	(b)
11.	(b)	12.	(c)	13.	(b)	14.	(c)	15.	(d)	16.	(a)	17.	(c)	18.	(a)	19.	(b)	20.	(c)

Short Answer Type

1. What are the typical characteristics of good building stones?
2. What do you mean by dressing of stones? State the advantages of dressing of stones.
3. Draw a neat, labelled diagram of cross-section showing structure of timber.
4. Discuss various defects that occur in timber.
5. How is bitumen produced? Explain various types of bitumen and their usage.
6. "Lime is a basis construction material", Justify the statement stating the properties of lime.
7. State the applicability of various types of soil in construction.
8. What is sand? What are the properties of good sand?
9. Classify coarse aggregates based on shape.
10. Write a short note on light weight aggregates.

Long Answer Type Questions

11. Where do you obtain building stones from? Discuss various methods of quarrying of stones.
12. What are the tools used for dressing of stones? Draw diagrams of these tools.
13. What is the purpose of seasoning of timber? Explain various methods of seasoning stating their relative merits and limitations.
14. What are coarse aggregates? Why coarse aggregates are added to make concrete? How do you classify coarse aggregate based on size?
15. Explain various types of limes and their usage in construction.

PRACTICAL

Experiment-1

Aim: Identify various sizes of available coarse aggregates from sample of 10 kg in laboratory and prepare report (60, 40, 20, 10 mm).

Materials Required: Sample of coarse aggregates (10 kg).

Apparatus: IS Sieve Set (as per IS 2386-1 1963), Weighing Balance (with accuracy to 0.1% of weight of sample), Tray, Mechanical Sieve Shaker (optional) and Oven.

Theory: This experimentation is conducted by dry sieve method to know the particle size distribution of coarse aggregates. The aim of the practical is to determine the grading of given coarse aggregates to assess their suitability for the proposed construction work.

Test Sample Preparation: The sample is prepared from larger sample as per the table below:

Maximum Nominal Size of Aggregates (mm)	Approx Minimum weight of Sample (kg)
63	100
50	100
40	50
25	50
20	25
16	12
12.5	6
10	3

Procedure:

1. Take the given sample of coarse aggregates and dry it in oven to a constant weight at a temperature of 100°C plus or minus 5°C.
2. Determine the weight of the dried sample accurately.
3. Arrange the sieves of IS sieve set in the order of decreasing opening size from top to bottom. Put the sample or a portion of the sample in the topmost sieve. Agitate the sieves manually or by mechanical shaker for a sufficient specified time period.
4. Weigh the mass of residue remained on each sieve very carefully. Record the readings in the report.

Observations and Calculations

Sample Wight = ...kg (original)

Weight of dried sample = ...kg

Sieve Size (mm)	Weight Retained (grams)	% Weight Retained	Cumulative % Weight Retained	% Passing
80				
40				
20				
10				
4.75				
Pan				

Result:

$$\text{Fineness Modulus of Coarse Aggregate} = \frac{\text{\% cumulative weight retained on sieve size (between 80 – 4.75)}}{100} + 500$$

Conclusion/Report: Report the fineness modulus to nearest 0.01.

Precautious:

1. The weight of dried sample should not differ the weight of original sample by more than 0.3%.
2. The sieving should be continued for a sufficient time period, say 15 minutes for manual operation and 5 minutes for mechanical shaker.
3. The weighing balance should be accurate.
4. The sieves used should be clean and free from any deposition.

Experiment-2

Aim: Identify the grain distribution pattern in given sample of teak wood in the laboratory and draw the various patterns.

Materials Required: Sample pieces of teak wood, Magnifying Lens.

Theory: Teak is one of the most superior types of timber used for making doors, windows frames and furniture. The advantages of teak wood are– it is hard, good quality, longer life span, easy to work with, having natural durability and dimensional stability. Indonesian teak (from the island of Java) is considered the most beautiful variant of teak. In modern times, teak is obtained from forests of Madhya Pradesh (India), Africa or Central America.

The quality of teak varies in the same timber in terms of its colour, grains, weight and oil content. These properties need to be assessed to estimate the qualities of teak like its fire resistance, resistance to attack by pests and environmental agencies.

- The grain of teak is straight or mostly straight. Also, the teak wood is darker at grains than rest of the wood. If the grains are not straight, then the teak under consideration is questionable.
- The teak tree grows in tropical climate only. During these conditions, the teak wood becomes hard and stable and develops naturally beautiful long straight grains.
- The grains are coarse and dark due to varied lands of silica.
- The teak can also be adjudged by touch. If you rub your hands on the teak wood, it will not shatter. Teak wood is very smooth to touch.

Procedure:

1. Prepare sample of wood from the given log of teak wood in definite rectangular shape and size.
2. Carefully observe the prepared sample for grain distribution patterns.
3. If the patterns are not clear, use magnifying glass to see the patterns along length and also in transverse direction.
4. Record the patterns by drawing.

Observations:

Sample No.	Grain Pattern (Straight/Mostly Straight/ Wavy/Irregular)	Grain Colour (Dark, Medium Dark, Light)	Remarks
1			
2			
3			

Sample No.	Grain Pattern (Straight/Mostly Straight/ Wavy/Irregular)	Grain Colour (Dark, Medium Dark, Light)	Remarks
4			
5			

Result: Coarse Dark coloured straight grains indicate good teak wood.

Precautions:

1. The samples should be prepared with most care with no harm to natural grain patterns.
2. As the observations are subjective in nature, it is importance to apply intelligence while observing grain patterns.
3. If grains pattern is not clear, wherever necessary, additional tests like touch, density and hardness should be performed on teak wood.

Experiment-3

Aim: Prepare the lime putty by mixing lime (1 kg) with water in appropriate proportion and prepare report on slaking of lime.

Materials Required: Lime i.e., calcium oxide, water.

Apparatus Required: Weighing Balance, Mixing Glass Jar, Mixing Wooden Rod.

Theory: Lime is chemically referred as calcium oxide. It is also called quick lime. When water in limited quantity is mixed with quick lime, the following chemical reaction takes place–

$$CaO + H_2O \rightarrow Ca\,(OH)_2 \text{ (exothermic)}$$

This process of making calcium hydroxide from lime is called slaking of lime. It is a white amorphous powder. Such hydraulic lime is commonly known as slaked lime or caustic lime. It is a white powdery compound which can also form colourless crystals. This reaction produces a lot of heat with hissing sound. The addition of limited amount of water breaks the lumps of lime and increases the surface area.

The hydraulics or slaked lime crystals/powder when suspended with water, results in a milky product which is used in preparation of lime putty for white washing. If dry slaked lime is passed by action of chlorine, it results in bleaching powder used for purification of sugar. If carbon dioxide is passed through aqueous slaked lime solution, lime water is produced. Slaked lime is used in construction projects for plaster, whitewash and as mortar.

Procedure:

1. Take 1 kg of pure lime i.e., quick lime (CaO) which comes in pieces and weigh it accurately.
2. Place lime pieces in a big tank and add some water in it.
3. The chemical reaction i.e., slaking of lime has started. It is an exothermic reaction with hissing sound.
4. Stir the mixture toughly and carefully with a spade.
5. Leave the lime well submerged in water for two weeks. Keep a regular check on tank to make sure that water is not all used up and add some more water, if necessary. Slaking of lime sometimes takes even months.
6. Keep on stirring the mixture on daily basis.

7. Continue the exercise, till a paste like product is obtained. It is lime putty which can be used for plastering purpose.

Observations:

Sample No.	Weight of Lime taken (kg)	Water Added (Ltrs.)	Time Taken to complete slaking to get lime putty (Days)
1			
2			
3			
4			
5			

Result: The longer you slake the lime, the better lime putty is achieved.

Precautions:

1. The addition of water to quick lime is a bubbly exothermic reaction. Care should be taken to avoid any burn by splash.
2. Gloves, Aprons and Goggles should be used while preparing putty.
3. The mix should be stirred properly to avoid its hardening. Stirring brings back plasticity in the mix.

Experiment-4

Aim: Identify various layers and types of soil in foundation pit by visiting at least 3 construction sites in different locations of city and prepare report consisting of photographs & samples.

Apparatus Required: Digging Tools, Sampler Jars.

Visit: Visit to 3-6 construction sites at foundation level situated in different locations of the city.

Theory: The soils under the ground vary in their physical properties with depth of excavation. It is possible to have varied kinds of soil in various layers in the same foundation pit. Two most important physical properties of soil are texture and structure. The texture of soil varies with the varying proportions of sand, silt and clay. This affects the water carrying capacity of soil and its porosity. These properties, in turn, affect the bearing capacity of soil. Hence, it is important to judge the nature of soil at a given site before finalizing the type of foundation for the proposed structure there. Further, it is important to note that the physical inspection of soil is only a preliminary estimate of its properties and the exact nature/ type of soil can be determined through particle size distribution (PSD) by sieve analysis in laboratory. The load bearing capacity of soil is another significant test to know the strength of a given site area for construction.

Procedure:

1. Visit the construction site with digging tools and sampler jars.
2. Examine various layers of soil in the foundation pit. The layers differ in colour and texture.
3. Take out soil sample from different layers to perform simple field tests to known soil type.

4. Perform following tests on soil sample:
   (a) **Visual inspection:** Try to assess the size and colour of the particles of soil. Soil particles larger than 4.75 mm and smaller than 80 mm are called gravel; soil particles ranging in size from 4.75 mm down to 0.075 mm are called sand. Fine sand cannot be easily distinguished from silt. Silt looks a little darker in colour.
   (b) **Dispersion Test:** Pour a spoonful of samples in a jar of water. If the material is sand, it will settle down in one minute or two minutes, but if it is silt, it may take 15 minutes to 1 hour.
   (c) **Shaking Test:** In this test, a part of soil mixed with water to a very soft consistency is shaken after placing it in the palm of the hand. If the soil is silt, water will rise quickly to the surface and give it a shiny glistening appearance. If it is clay, the water cannot move easily and hence it continues to look dark.
   (d) **Dry Strength:** The strength of soil in a dry state can be estimated by crushing it between thumb and forefinger. A silt fragment crushes easily whereas a clay fragment cannot be crushed easily.
   (d) **Rolling Test:** Attempt to make a thread out of a moist soil sample with a diameter of about 3mm. if the material is silt, it is not possible to make such a thread without disintegration and crumbling. If it is clay, such a thread can be made even to a length of about 30 cm and supported by its own weight while holding by ends.

Observations:

Sample No.	Layer No.	Size	Colour	Time for Settlement (Minutes)	Type of soil (Sand/Silt/Clay)
Sample 1	Layer 1				
	Layer 2				
	Layer 3				
Sample 2	Layer 1				
	Layer 2				
	Layer 3				
Sample 3	Layer 1				
	Layer 2				
	Layer 3				

Result: The soil sample belongs to .... category.

Precautions:

1. Take out the soil sample carefully without disturbing the adjacent layers.
2. Take at least 3 samples from foundation pit of one construction site.

KNOW MORE

1. **Soil Suitability:**

Sr. No.	Type	Usage
01.	Gravel	No usage for wall building. Used as a filler in construction
02.	Sand	Suitable of foundation surface. Used in wall construction only when mixed with clay.
03.	Silt	Best foundation surface. Not suitable for wall or pressed block building.
04.	Clay	Not suitable for foundation surface.
05.	Organic Soil	Useless as a building construction material.

2. **Stabilizers for Soil:**

- Straw
- Plant Juice
- Cow Dung
- Flyash etc.
- Cement
- Molasses
- Oil

3. **Interesting facts on Teak:** Teak is one of the most superior types of timber used for making doors, windows frames and furniture. The advantages of teak wood are- it is hard, good quality, longer life span, easy to work with, having natural durability and dimensional stability. Indonesian teak (from the island of Java) is considered the most beautiful variant of teak. In modern times, teak is obtained from forests of Madhya Pradesh (India), Africa or Central America. The quality of teak varies in the same timber in terms of its colour, grains, weight and oil content. These properties need to be assessed to estimate the qualities of teak like its fire resistance, resistance to attack by pests and environmental agencies.

REFERENCES AND SUGGESTED READINGS

- Duggal, S.K., Building Materials, New International, New Delhi.
- Arora S.P and Bindra S.P., Building Construction, Dhanpat Rai Publication, New Delhi.

INTERESTING FACTS

The most interesting fact about almost all natural construction materials in that their usage is eco-sustainable and easily re-usable. The natural building materials are low cost and flexible in all phases of use. In short, the most of the natural building materials are anti seismic, fire resistance and provides a reasonable good shield against environmental factors like temperature, humidity and radiations. Besides the discussed natural materials, there are many more naturally materials which have been used in history for construction. These include:

(i) **Cob:** A mix of subsoil, water and fibrous content like straw.

(ii) **Sheep Wool:** It can be used as fibre in earth work at ceilings, walls etc. to control temperature at cold regions.

(iii) **Recycled Wood:** Instead of looking for fresh timber, recycled wood can be used in aesthetically pleasing floors, exposed beams and other indoor wooden works.

(iv) **Cork:** Cork is a natural wooden product obtained from oak tree. It is beneficially used in wall panels to maintain desired acoustics and absorb vibration. It is resistant to moisture and any other liquid also.

(v) **Straw:** Just like wood, straw is used in walls and ceilings to maintain temperature stability of the structure. Straw can be compressed to make cladding which can be installed in structures. These are recyclable and biodegradable.

(vi) **Ash:** Various types of naturally producing ashes like Flyash, Bottom Ash, Rice Husk Ash and many more may be used in concrete as mineral admixture to get some desired properties like increase in workability, smaller pores and ultimately better strength and durability.

(vii) **Hampcrete:** It is a mixture of sand, hemp fibres and lime. The building unit blocks made up of hampcrete are light in weight and have better thermal insulation. The use of hampcrete is of size in construction industry.

Some examples of natural construction materials projects are given in enclosed QR.

3 Artificial Construction Materials

UNIT SPECIFICS

This unit specifies the following aspects:

- Knowing various artificial construction materials.
- Understanding the use of appropriate artificial building material for specific purposes.
- Knowing the detailed procedure for manufacturing of various artificial construction materials.
- Understanding subcategories or types of various artificial construction materials.
- Gaining practical exposure in identification of quality of various artificial construction materials.

Besides, the above aspects, the experimental consideration to understand various artificial construction materials viz their types, field uses, and quality check has also been explained in this chapter. This will enrich the young minds on the appropriate selection procedure for various materials for specific purpose. Students will develop an inquisitive approach towards performing field tests on construction materials while selecting these for a particular project.

A number of objective multiple choice questions have been given at the end of the unit to prepare students for competitive exams. Besides, short, and long answer type questions have been framed based on bloom's taxonomy to assess the level of understanding for students. Further, know more section and suggested readings have been shared to gain some more knowledge on artificial construction materials and their applications. Some QR Codes have been put in the unit to get deeper knowledge of artificial construction materials.

At the end of this chapter, practical's related to classification of bricks, quality of bricks, and identification of various types of flooring and that of glasses have been covered, based on which students can actually perform these test in the field. The whole chapter has been prepared with utmost care to ensure good understanding of artificial construction materials.

RATIONALE

The unit on artificial construction materials deals with the manufacturing and applications of various man made building materials. The basic constituents of these artificially made materials, classification of these materials, their manufacturing methods, field tests to analyse their quality and the related specific uses and applications have been discussed in great details. This chapter provides technical exposure to users on variety of artificial construction materials like cement, bricks, tiles, precast blocks, plywood, veneer, glass and metals. Though a lot of naturally occurring materials are being used in construction purpose, artificially made construction materials have their own importance in infrastructural projects. A rationale on the balanced use of natural and artificial construction materials can be framed after understanding this chapter thoroughly.

PRE-REQUISITE

Basic knowledge on construction materials (i.e., Unit-I)

UNIT OUTCOMES

Following is the list of outcomes from unit on Artificial Construction Materials:

U3-O1: Define and understand various artificial construction materials.

U3-O2: Understand the desired properties of artificial construction materials.

U3-O3: Understand the manufacturing process of various artificial construction materials.

U3-O4: Know the uses of each artificial construction material.

U3-O5: Apply the knowledge gained to select appropriate artificial construction material for a specific purpose.

Unit-3 Outcomes	EXPECTED MAPPING WITH COURSE OUTCOMES (1- Weak Correlation; 2- Medium Correlation; 3- Strong Correlation)				
	CO-1	CO-2	CO-3	CO-4	CO-5
U3-O1	3	3	1	2	2
U3-O2	2	3	1	1	1
U3-O3	2	3	1	2	1
U3-O4	3	3	1	2	1
U3-O5	3	3	1	3	1

3.1 BRICKS

Brick is a single block made up of clay or ceramic and used in masonry construction work. The bricks are laid in brick work and are held together using mortar to make a permanent structure. Bricks, as building material, are used in construction of load bearing walls, pavements and as non structural members for partitions etc. Bricks are either of uniform size as specified by Bureau of Indian Standards i.e., 190 mm × 90 mm × 90 mm or of convention size of 230 mm × 114 mm × 75 mm. The convention size may vary from region to region in India.

3.1.1 Constituents of Good Brick Earth

The good brick earth mainly constitutes of clay and sand i.e. alumina and silica besides, some other constructions are also either present or added in brick earth to get some desired properties like ease of drying without shrinkage, cracking or warping. These additional ingredients include lime (Chalk), magnesia, oxide of iron etc. in small quantities. The functions of each constituent of brick earth are as explained follows:

(i) **Clay (Alumina):** Alumina, a major constituent of any type of clay, absorbs water and makes the clay plastic so that it can be easily moulded into desired shape and size. A good brick should contain 20 – 30% of alumina. It is further important to note that excess of alumina in clay may make the brick liable to shrink or warp during drying or burning process.

(ii) **Sand (Silica):** Silica present in clay or sand, when combines with alumina, it forms silicate of alumina. It prevents the bricks from shrinkage, cracking or warping. The quantity of silica in a good brick earth is generally between 50 – 60%. Excess of silica in earth may destroy cohesion property and it may make bricks too brittle.

(iii) **Lime:** Lime, in small quantities (not more than 5%) in brick earth is desirable. It controls the shrinkage of raw bricks and helps in proper binding during burning. It should be present in brick earth in fine powdery form and not as lumps. The excess of lime may cause brick to lose its shape due to melting during burning.

(iv) **Iron Oxide:** Iron oxide imparts red colour to bricks. Its presence in small fractions (5 – 6%) helps the sand to fuse together during burning and gain good strength and hardness. However, the excess of oxides of iron may impart dark blue colour to bricks.

(v) **Magnesia:** A very small quantity of magnesia ($\approx$1%) imparts yellowish tint to the bricks. Presence of magnesia prevents shrinkage but, if present in excess, magnesia may decay bricks.

Besides, these main constituents, some harmful ingredients may be present in bricks earth. These include– Alkali, whose presence in excess, results in flux during burning of bricks causing de-shaping. Further, if alkali rich bricks are used in brick masonry, these absorb moisture from the atmosphere and when this moisture evaporates due to sunlight, a layer of white or grey deposits is seen on the surface. It is called 'Efflorescence'. Also, the presence of gravels or pebbles is undesirable in brick earth as their presence results in cracks in bricks on drying. The presence of organic matter in brick earth is also undesirable as it makes the bricks porous in nature.

3.1.2 Conventional/Traditional Bricks

Conventional bricks are those which have been used in different regions/places from time to time and which have not been standardized in size. The dimensions of traditional bricks vary i.e., length between 20 – 25 cm, width between 10 – 13 cm and thickness varies between 3 – 7.5 cm. The most commonly adopted size of traditional bricks has been 23 cm × 11.4 cm × 7.6 cm approximately.

3.1.3 Modular/Standard Bricks

For the uniformity of size of bricks, Bureau of Indian Standards (BIS) has laid down specifications for dimensions of bricks. These are 19 cm × 9 cm × 9 cm. The bricks which conform to the standard size laid down by BIS are called modular or standard bricks. Modular bricks have their own advantages like economy in manufacturing with repeated usage of moulds, requirement of less area for drying and staking and pleasing look of exposed surface. In general, the masonry work with standard bricks is economical than that with conventional bricks.

3.1.4 Special Bricks

A few types of special bricks are also in use in modern construction era. These bricks are made either in special shapes to give the desired look to the surface from architectural point of view or are made of some special ingredients other than normal constituents of brick earth. These bricks include:

(i) **Concrete Bricks:** These are made up of sand, cement, aggregates and water. These are customized in different sizes and desired shapes and can be manufactured at site itself.

(ii) **Lime Bricks:** The brick made up of sand and limestone is called lime bricks. These improve the visual aesthetics of the structure.

(iii) **Engineered Bricks:** The bricks which are engineered by adjusting the proportion of its ingredients to get increased strength and better resistance to natural attacking agencies like fire, wind and water. These bricks are mostly used in basements and foundations i.e., the areas which are more likely to get exposed to moisture.

(iv) **Fly ash Bricks:** These bricks are made up of fly ash and water. Fly ash bricks have the advantage of better resistance to freeze and thaw conditions. These bricks are light in weight (approximately 40% weight of conventional bricks) but strong enough. Fly ash bricks have low water penetration and these have better insulation. The use of fly ash bricks is on rise because of their economy and overall better performance. The composition of fly ash bricks may be kept as flyash 50 – 60%, sand 20 – 30% and lime 10 – 20%. Further, gypsum in small proportion is added to act as long term strength gainer. Fly ash bricks have other advantages like these do not require soaking in water before usage and, practically, there is no breakage of bricks during transport and use. However, there are some disadvantages also associated with the use of fly ash bricks. These are early low strength as the rate of gain of strength is slow in fly ash bricks and these bricks have high thermal conductivity.

3.1.5 Characteristics of a Good Brick

A good standard brick should possess following characteristics:

- Bricks should have sharp edges and uniform in size and shape.
- A good brick should be sound proof.
- A good brick should not break into pieces when it is dropped from a height of about 1 meter.
- Bricks should have low thermal conductivity.
- Bricks when broken should have a homogenous, uniform compact structure, free from any voids.
- Bricks shall have a crushing strength not less than 3.5 N/mm^2.
- Good bricks should have good fire resistance.
- Good bricks should not have water absorption more than 20%.

3.1.6 Field Tests on Bricks

- **Appearance:** The bricks should have uniform crimson red colour. It should have sharp edges.
- **Strength:** A good brick should not break into pieces when it is dropped from a height of about 1 meter.
- **Hardness:** When a brick is stretched with a finger nail, it shall not leave any impression. It indicated good hardness of the brick.
- **Compactness:** Bricks when broken should have a homogenous, uniform compact structure, free from any voids.
- **Soundness:** When two randomly selected bricks are struck with each other, it should give a clear raining sound and the bricks should not break.

- **Soluble Salts:** The bricks should be free from any efflorescence i.e., any white powder on its surface. Presence of efflorescence shows presence of soluble salts in bricks. For this, immerse the brick in water for 24 hours and then dry it in shade. The dried brick should not have any deposits on its surface.

3.1.7 Aerated Concrete Blocks

Aerated Concrete Blocks (ACB) are a good substitute to traditional usage of bricks. These blocks constitute of fly ash, lime, cement, gypsum and water. The concept of aerated concrete blocks was given in 1924 by a Swedish architect. The processes of mixing and the proportion of raw materials decide the properties of these blocks. The composition of aerated concrete blocks may be fly ash (60 – 70%), lime (10 – 20%), cement (10 – 20%), gypsum (3 – 5%) with a water/dry material ratio as 0.6 – 0.65. As these blocks have a lot of air aerated during their manufacturing, the blocks are very light in weight (say density 550 – 650kg/m^3). The size of these blocks can be adjusted as per need, however, the general size of aerated concrete blocks in 600 mm × 200 mm with thickness varying between 75 mm to 300 mm.

The aerated nature of blocks provides good thermal insulation, fire resistance and sound absorption. Hence, aerated concrete blocks are considered as an eco-friendly green building material having good properties of strength and durability.

3.1.7.1 Advantages of Aerated Concrete Blocks

(i) The use of an industrial by-product i.e., fly ash makes ACB units an eco-friendly and sustainable building block.

(ii) ACB units are 3 – 4 times lighter than normal bricks and hence reduce the dead load of the structure.

(iii) The air pores present in AC blocks, provide very good thermal insulation and sound absorption property.

(iv) Aerated blocks are easy to be cut, drilled, nailed and grooved as and when required.

(v) The use of ACB units reduces the construction time by 20% as lighter units are easy to transport and handle.

3.1.7.2 Limitation of Aerated Concrete Blocks

- AC blocks are brittle in nature and needs to be handled with care to avoid breakage.
- As AC blocks are thicker in size, they leave less area for usage in rooms when walls are made up of AC blocks.
- Aerated blocks are difficult to manufacture in rainy season as proper drying cannot be achieved.

3.1.7.3 Comparison of Bricks and Aerated Concrete Blocks

Table 3.1: Comparison of Bricks and Aerated Concrete Blocks

Parameter	Brick	Aerated Concrete Block
Cost	–	Results in a saving of 15%
Plastering	Required more mortar for plastering due to irregular surface and more number of joints	Requires less plastering as the surface are smother with less joints.

Parameter	Brick	Aerated Concrete Block
Quality	Varying in nature	Uniform quality
Speed of construction	Slow	Relatively fast as these are light in weight
Breakage	10 – 20% average	5 – 7% average
Energy Consumption	No Saving	30% reduction in energy consumption per structure due to better insulation
Acoustics	–	Better acoustics due to sound absorption
Availability	Throughout the year	Shortage in rainy session

3.1.8 Classification of Bricks

Bricks are burnt in kiln. Kiln burning has control on burning and hence, it gives good quality bricks with good strength, durability and hardness. The burnt clay bricks are classified into four types as follows:

3.1.8.1 First Class Bricks

These are the best quality bricks which are thoroughly and uniformly burnt. The characteristics of first class, also known as class A bricks, are as follows:

(a) The bricks are of uniform colour (Crimson Red) as these are thoroughly burnt.
(b) First class bricks have smooth rectangular faces with sharp corners and edges.
(c) These bricks are free from flaws and cracks.
(d) When two first class bricks are struck with each other, it gives a clear metallic ringing sound.
(e) A broken surface of a first class brick shows a uniform compact texture.
(f) Scratching of first class bricks with finger nails leaves no marks on its surface.
(g) These bricks are free from or have only slight efflorescence.
(h) First class bricks absorb not more than 10% water by their weight when immersed for 24 hours.
(i) A tolerance limit of 3% in compressive strength is permitted for first class bricks.
(j) There is no sign of presence of free lime in these bricks.

Uses: These are best quality bricks and their suitability at site is as follows:

(a) First class bricks are used in permanent structures with long durability requiring superior and sound works like monuments, heritage building and repair of historical buildings etc.
(b) These bricks are used in face works of structures which are not to be plastered.
(c) These bricks are also used as aggregates in flooring RCC works and in reinforced brick masonry work.
(d) These are used for buildings which are exposed to corrosive environments.

3.1.8.2 Second Class Bricks

Second class bricks, also called class B bricks, are very similar to first class bricks. These are also hard and well burnt bricks but there may be some irregularity in shape and size of these bricks. The characteristics of second class bricks are quite same as that of first class bricks and there is no difficulty in laying the bricks but the surface of these bricks may be somewhat rough. Second class bricks exhibit properties similar to first class bricks in terms of strength and durability but the aesthetic appeal is less for second class bricks. These are:

(a) used in structure where the surface is to be plastered.
(b) used as bricks ballast in RCC structures.
(c) used in temporary sheds buildings where life intended is not more than 15 years.

3.1.8.3 Third Class Bricks

Also known as class C bricks, third class bricks have following characteristics:

(a) These are slightly under burnt bricks and hence are lighter in their red colour.
(b) The shape and size of these bricks are irregular and hence there is difficulty in laying these bricks.
(c) Third class bricks are soft and have less crushing strength.
(d) These bricks give dull or blunt sound when struck with each other.
(e) The surface of these under burnt bricks shows considerable efflorescence.
(f) When these bricks are stretched with finger nails, an expression is clearly left on the surface.
(g) The water absorption capacity of these bricks is between 15 – 20% of their weight in 24 hours.

Uses: Third class bricks are not suitable for brick masonry works. These are inferior bricks in terms of strength, durability and aesthetic sense and hence are not preferred. However, sometimes these bricks are used as filler material under RCC footing and flooring.

3.1.8.4 Over burnt or Jhama Brick

These bricks are also known as vitrified bricks as these bricks get over burnt in kiln either due to being near to the fire in kiln or being exposed to fire for a longer period. Due to over burning, the bricks get fused and lose their shapes, size and sharp corners. These bricks have distorted shape and have a high water absorption capacity:

Uses: Over burnt bricks are not suitable for brick masonry works or exposed surface due to difficulty in laying. These are:

(a) Used as coarse aggregates in concrete.
(b) Used in the foundation of structures.
(c) Used as road metal aggregates.

3.1.8.5 Comparison of Various Types of Burnt Bricks

Table 3.2: Comparison of various types of burnt bricks

Property	1st Class Bricks	2nd Class Bricks	3rd Class Bricks	Over burnt Bricks
Edges and Corner	Sharp	Rounded	Irregular	Fused
Surface	Smooth	Rough	Very Rough	Coarser
Colour	Crimson Red	Yellowish Red	Light Red	Dark/Blackish Red
Sound	Metallic Sound	Metallic Sound	Dull Sound	Blunt Sound
Burning Status	Thoroughly brunt	Uniformly burnt	Under burnt	Over burnt

Property	1st Class Bricks	2nd Class Bricks	3rd Class Bricks	Over burnt Bricks
Tolerance Limit in Crushing Strength	3%	8%	10%	–
Efflorescence	Nil	Slight	Considerable	Too high
Water Absorption	<10%	<15%	15 – 20%	>20%

3.1.9 Manufacturing Process of Burnt Clay Bricks

The manufacturing of conventional bricks is carried out in four distinct steps. These are preparation of clay, moulding, drying and burning.

3.1.9.1 Preparation of Clay

The clay or earth used for manufacturing of bricks needs to be prepared well. There are six basic stages involved in the preparation of clay or brick earth. These include:

(a) **Un-soiling:** This operation involves removal of the top soil from the site up to a depth of approximately 200 mm. This layer is removed as it contains a lot of impurities and organic matter.

(b) **Digging:** The clay is then dug out and spread on a level ground, a little below the general ground level in the form of small heaps of height 60 – 120 cm.

(c) **Cleaning:** In this step, the dug clay is cleaned from impurities like stone, pebbles, vegetable matter etc. Lumps, if any, are reduced to powder form. However, if the impurities content is high, the clay should be washed and screened. This makes the whole operation costly and such clay is not used in making of bricks.

(d) **Weathering:** The cleaned dug soil is exposed to atmosphere for softening. The process may take few weeks to few months depending up on the nature of soil. This operation improves the strength and plasticity of clay.

(e) **Blending:** The intense mixing of clay, sand and other materials of good brick earth is called blending. The process is carried out by taking small portions every time and turning it up and down to ensure uniform mixing.

(f) **Tempering:** The properly blended clay is mixed with water (approx. 25 – 30%) and kneaded/pressed under the feet of men or cattle till a homogeneous mass is achieved. This mass is covered with mats and then allowed to dry at a slow rate. This process softens the clay mix and gains plasticity and makes it fit for moulding.

3.1.9.2 Moulding of Bricks

The process of making rectangular units of standard size from the prepared tempered clay is called moulding. The bricks are moulded by following two methods:

(a) **Hand Moulding:** The moulding of bricks with hand is the most popular method in India. The rectangular mould of well seasoned wood or steel, open at top and bottom, are filled with prepared clay forcefully, to ensure that it fills all the corners. The extra clay is removed either by a wooden strike or an outing frame with wire or a metallic strike. Hand moulding is carried out either on levelled ground or on a standard table top and is accordingly referred as ground moulding or table moulding. Hand moulding is easy to carry out and is preferred where manpower is cheap and readily available. After filling the moulds, the mould/boxes are lifted up and the raw bricks are left for drying. When the bricks are sufficiently dried, these are carried to drying sheds and stacked properly. A depression, called frog, is made on the top surface of the brick to serve two purpose.

(i) To indicate trade name of the manufacture.

(ii) To serve as a key for mortar when bricks are placed over it.

(b) **Machine Moulding:** Machine moulding is carried out by using machines and is adopted when the bricks are to be produced in very large numbers. Two types of methods and machines are used for machine moulding of bricks:

(i) **Plastic Clay Machines:** In this method, the prepared clay, in plastic state, is forced into the openings of size equal to length and breadth of a brick in the machine and then cut into strips of thickness of brick with the help of wire frames and then dried as usual.

(ii) **Dry Clay Machine:** In this method, dry prepared clay is reduced to powder and filled into the rectangular moulds of the machine and is subjected to high pressure to form well shaped hard bricks, such bricks are called pressed bricks. The machine moulded bricks are superior to hand moulded bricks in terms of regular shape, sharp corners, heavier in weight and stronger.

3.1.9.3 Drying

Drying of moulded bricks is necessary as these cannot be taken directly for burning. Drying is basically carried out to:

(i) Increase the strength of raw brick so that it can be handled and stacked in kiln for burning without getting damaged.

(ii) Remove excessive moisture from the brick before burning to save fuel and burning time. Generally, the bricks are dried to 5 – 7% moisture content before burning.

Drying of bricks is carried out either naturally or by artificial means.

In natural drying, moulded bricks are placed in rows along and across in alternate layers, ensuring free circulation of air all around, under the sheds. The sheds protect bricks from bad weather and rains. The natural drying takes about 7 – 14 days for bringing down the moisture content to 5 – 7%.

Artificial drying is a faster method for drying bricks on a large scale. Special dryers, producing heat for drying are used in artificial drying.

3.1.9.4 Burning

Burning of bricks is the most important step involved in manufacturing of bricks. Burning impacts hardness, compactness, durability and resistance to decay of the bricks. Bricks should be properly and uniformly burnt as under burnt bricks are soft and are less strong whereas over burnt bricks lose their shape and size on melting and become brittle.

Burning of bricks mainly involve three reactions:

Dehydration (removal of moisture from brick earth at a temperature of 425 – 750°C)

Oxidation (elimination of carbon and sulphur as oxides) and oxidization of fluxes (at a temperature of about 900°C)

Vitrification (hardening of bricks at temperature beyond 900°C).

Bricks are burnt either in a clamp or in a kiln. Clamp is a temporary structure whereas kiln is a permanent structure. Clamp burning is a slow process carried out for small scale production of bricks; however, the bricks produced are cheap due to less consumption of fuel and hard due to slow burning. Kiln burning has higher initial and running cost, but it is more suitable for large scale production of bricks at a faster rate.

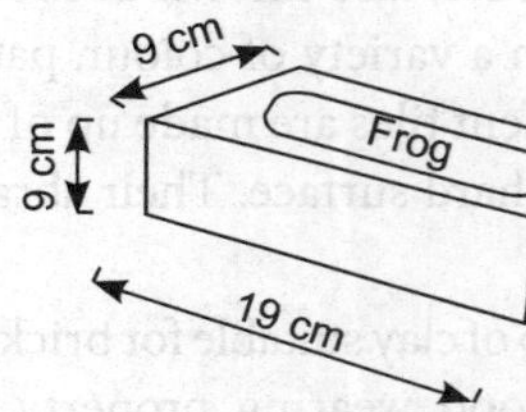

Figure 3.1: A Standard Brick

3.2 TILES

Tiles in building construction are thin plate like elements used to cover surface like flooring, roofs or even walls. The tiles are used for interior as well as exterior finishing. Tiles are generally made up of clay in powder form free from organic matter. The clay is pressed in regular modules, dried and subsequently burnt to obtain tiles. A large number of recycled materials are also been used these days to manufacture tiles. Tiles are available in glazed or unglazed forms. Based on the applications, tiles are classified as– Roofing Tiles, Flooring Tiles, Wall Tiles, Partition Tiles.

Here, we are discussing flooring tiles.

3.2.1 Flooring Tiles

The tiles which are used to make/ cover floors are called flooring tiles. Flooring tiles must give a smooth, durable, strong and abrasion resistance surface finish. The flooring tiles are designed to enhance the aesthetic look of the room, ensuring the comfort of movement. Flooring tiles can be placed both in interior as well exterior of the building. Flooring tiles are subjected to higher loading and pressure as compared to wall tiles and hence needs to be made stronger by burning at a higher temperature. Besides, flooring tiles should be easy to clean and impervious to water flow.

3.2.2 Types of Flooring Tiles

Based on the material, flooring tiles are classified as:

(i) **Ceramic Tiles:** These are made up of a mixture of clay and other materials. These are pressed at room temperature and then burnt in kiln. Some examples of ceramic tiles are earthenware tiles, terracotta, fully vitrified tiles etc. vitrified ceramic tiles are made from clay and other finds ground materials, burnt at very high temperature to ensure burning of complete thickness of tile. Vitrified tiles have better abrasion resistance and are as hard as granite. These are very commonly used for flooring purpose.

(ii) **Porcelain Tiles:** Porcelain tiles are also made up of clay but they are made by heavy and denser clay. These tiles are burnt at high temperature for a longer period i.e., till all the water present in material is evaporated. Porcelain tiles are harder and denser than ceramic ones. Though porcelain tiles are costlier, these are considered superior to ceramic tiles in terms of strength, durability, uniformity in design and colour and better abrasion resistance. Porcelain tiles have a lesser water absorption rate i.e., of the order of 0.5%. Porcelain tiles can be polished to get a glazed shiny look. These are preferred where there is heavy traffic. However, for residential building ceramic tiles can also serve the purpose suitably.

(iii) **Cement Floor Tiles:** Cement floor tiles, also known as mosaic tiles are also popular option for laying as floor. These are available in a variety of colour, pattern and provide a good alternative for decorative durable flooring. Cement tiles are made up of cement under good supervision and hence offer a dense, water tight and hard surface. Their abrasion resistance is better than cast in site cement floor.

(iv) **Quarry Tiles:** These tiles are made up of clay suitable for brick manufacturing. Quarry tiles are dense and have good wearing property. These tiles are preferred at places where floor is expected to be exposed to acid or alkali attack i.e., places like laboratories, chemical industries, paper industries, dyeing and textile industries, tanneries etc.

Besides, flooring tiles can be classified on the basis of their surface finish either as glossy surface or matt finish tiles having rough surface to avoid slippery action on wet floor. Matt finish tiles are used in bathrooms and toilets for this reason.

3.3 CEMENT

Cement is the basic ingredient of a concrete and is one of the most commonly used artificial construction material. The most popular type of cement is Ordinary Portland Cement (OPC) which consists of three ingredients namely lime, silica and alumina. A few other ingredients are also added in cement in small quantities like iron oxide, magnesia, alkalis and gypsum. The composition of OPC and the function of each ingredient are as under:

Table 3.3: Function of Each Ingredient of Cement

Ingredient	% Range	Function
Lime (CaO)	60 – 65	Gives Strength
Silica (SiO_2)	17 – 25	Gives Strength
Alumina (Al_2O_3)	4 – 8	Quick Setting Property
Iron Oxide (Fe_2O_3)	0.5 – 4	Impart Colour
Magnesium Oxide (MgO)	0.1 – 3	Impart Hardness and Colour
Sulphur Oxide (SO_3)	1 – 2	Soundness
Alkalis	0.3 – 1	Undesirable, Causes efflorescence
Impurities	0 – 0.5	Undesirable
Free Lime	0 – 1	As impurity

Besides, a small amount i.e., 3-5% of gypsum is added while grinding cement, which acts as a retarder and controls the setting time of cement.

3.3.1 Manufacturing of Cement

There are two methods for manufacturing of cement. These are

(i) Dry Process (ii) Wet Process

The basic operations involved in both the processes are as follows:

(i) Mixing of raw materials (ii) Burning

(iii) Grinding

However, the two methods have their own advantages and limitations. Dry process for manufacturing of cement is popular in United States of America and some other developed countries but in India, wet process is more popular for manufacturing of cement. The process of production of cement is relatively slow and tedious in dry process when compared to wet process. However, the advantage of dry process is that energy required to burn the ingredient is much less than the wet process. Hence, it is observed that dry process is confined to hard stone type material and at all other places, wet process is found to be more suitable. The flow diagrams of the two processed of manufacturing of cement is as follows:

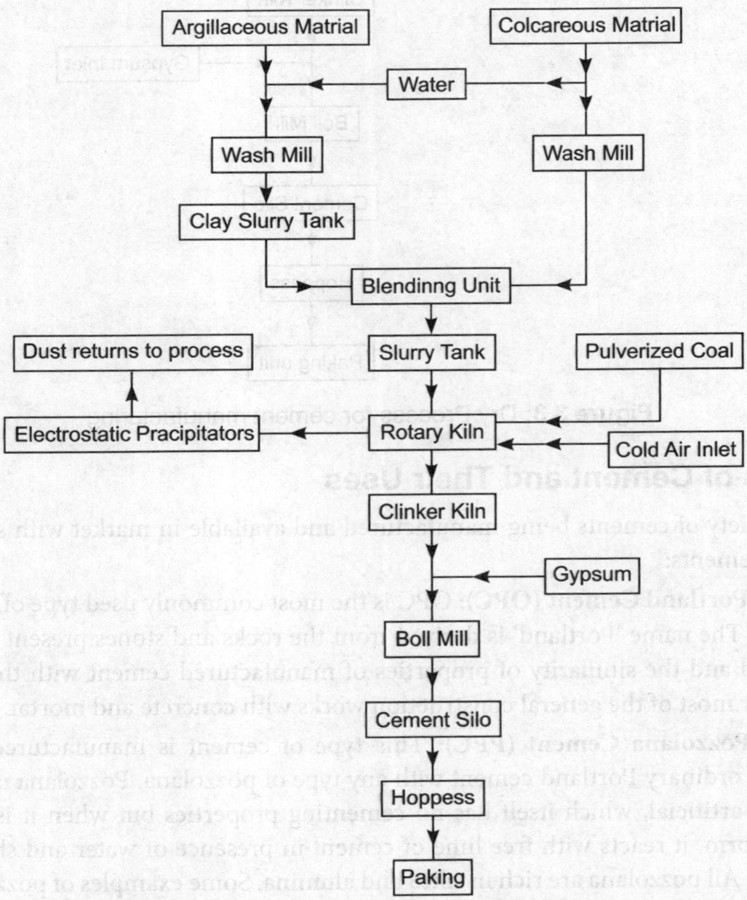

Figure 3.2: Wet Process for cement manufacturing

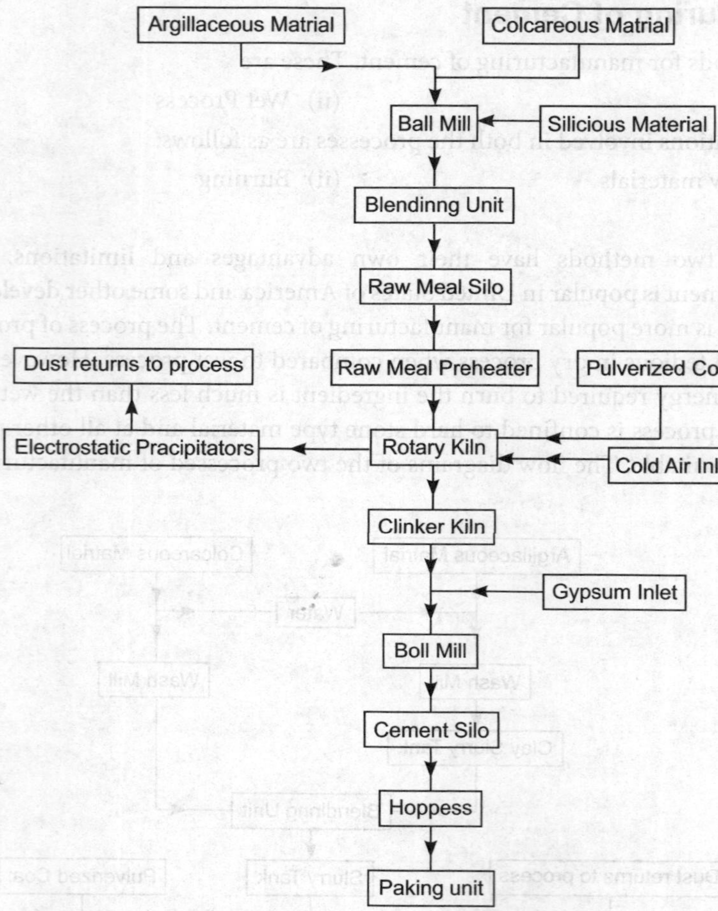

Figure 3.3: Dry Process for cement manufacturing

3.3.2 Types of Cement and Their Uses

There is a big variety of cements being manufactured and available in market with specific usage. The various types of cements:

(a) **Ordinary Portland Cement (OPC):** OPC is the most commonly used type of cement all around the world. The name 'Portland' is derived from the rocks and stones present in Isle of Portland in England and the similarity of properties of manufactured cement with those rocks. OPC is suitable for most of the general construction works with concrete and mortar.

(b) **Portland Pozzolana Cement (PPC):** This type of cement is manufactured by grinding the clinkers of ordinary Portland cement with any type of pozzolana. Pozzolana is a material, either natural or artificial, which itself has no cementing properties but when it is ground to a fine powdery form, it reacts with free lime of cement in presence of water and shows cementitious properties. All pozzolana are rich in silica and alumina. Some examples of pozzolana are volcanic ash, fly ash, blast furnace slag etc. PPC gains strength slowly and hence, its early strength is less than that of OPC but the ultimate strength is similar to ordinary Portland cement. Due to slow rate of hydration, PPC has low heat of evolution and hence it is useful for mass concreting projects like dams, bridges and under water projects. Portland pozzolana cement has a better resistance to sulphate and other chemical attack. The quantity of pozzolana in cement may vary between 15 – 50% by weight of cement depending on its usage.

(c) **Rapid Hardening Cement (RHC):** Rapid Hardening Cement is very similar to ordinary Portland cement in composition, but it gives strength much faster than OPC. Its 3 days compressive strength is comparable to 7 days strength of OPC. RHC is produced by adjusting the chemical composition in such a way so as to achieve a higher content of tri-calcium silicate (C_3S) and a finer grinding of cement clinker. This cement is used at places where removal of shuttering is required at an early age due to repeated casting requirements etc. Use of RHC results in considerable saving in time and money. The cement should not be confused with quick setting cement.

(d) **Quick Setting Cement:** This cement is often mistaken with rapid hardening cement. However, the two are different. Quick setting cement sets faster/quicker than OPC but its grain size and rate of gain of strength are similar to that of OPC. The initial setting time of quicker setting cement may be as low as 5 minutes against 30 minutes for OPC. This cement is useful for those locations which are exposed to stagnant or running water. Quick setting cement is not much popular because proper mixing and working with concrete in such a short duration is practically very challenging.

(e) **Low Heat Cement:** When the cement sets and hardens, a lot of heat is generated. In mass concreting construction projects like dams, bridges, abutment etc., this heat produced, causes shrinkage of concrete during curing and hence resulting in cracking of the concrete. To overcome this problem, low heat cement is used in mass concrete projects. Low heat cement is designed by keeping the content of C_3A and C_3S as low as possible as these compounds cause a lot of heat evolution. Also the cement is ground finer than OPC. Low heat cement has a longer initial setting time and also rate of gain of strength is slower. Hence, the strength of this cement is lower at initial stages but the ultimate strength is same as that of OPC.

(f) **Sulphate Resisting Cement:** Ordinary Cement has a tendency to react with sulphates present in soil or ground water and form sulpho-aluminates which are expansive in nature and cause disintegration of concrete. Sulphate resisting cement is produced by restricting the quantity of C_3A and C_4AF to less than 5%. This cement has strength similar to that of OPC. Sulphate resisting cement is beneficial under severe alkali conditions like for foundation under soil containing high content of sulphate and for canal lining, pipe lines, culverts etc.

(g) **Blast Furnace Slag Cement:** Slag is a by-product produced in the manufacture of pig iron. It contains a lot of lime, silica and alumina. Blast furnace slag cement is produced by grinding Portland cement clinker with granulated blast furnace slag. The amount of slag is generally kept between 25% – 65%. Blast slag cement is much more economical than OPC. This cement gains strength slowly and has low heat of hydration. This cement is useful for mass concreting projects in hot weather. This cement also has the advantage of more resistance to sulphates.

(h) **High Alumina Cement:** High alumina cement is a special kind of cement produced by melting bauxite and lime together and then grinding it with OPC clinkers. This cement has very high compressive strength. This cement is much more workable and is used for projects under extreme weather condition like high temperature and frost.

(i) **White Cement:** White cement is similar to OPC except that it is white in colour instead of grey. This is achieved by restricting the iron oxide content to less than 1%. This cement is expensive then OPC and hence, is used for aesthetically important and decorative projects only.

3.3.3 Field Tests on Cement

The following tests are prepared on cement in field or at site to assess its quality:

(i) **Date of Manufacturing:** The cement bag is checked to see the date of manufacturing as the strength of cement reduces with age due to some moisture absorption from atmosphere.

(ii) **Colour:** Cement should be of greenish grey colour and the colour of cement should be uniform throughout.

(iii) **No Lumps:** The cement should be free from any hard lumps as the presence of lumps indicates absorption of moisture from the atmosphere.

(iv) **Feel:** When the hand is immersed in to a bag of cement, it should give a cool feeling. If some hydration of cement has taken place, heat is generated and the feeling is warm.

(v) **Smoothness Test:** The cement particles when rubbed between the fingers, should give a smooth feeling as smooth as talcum powder.

(vi) **Sink Test:** If a handful of cement is thrown in to a bucket of water, it should float for sometime before sinking.

(vii) **Glass Plate Test:** A thick paste of cement and water is made on a glass plate and kept under water for 24 hours. On taking out, it should set and should not crack.

3.4 CONCRETE BLOCKS

The concrete blocks are the precast elements which are formed and hardened before they are brought to the site. The concrete blocks are manufactured in a factory under controlled supervision ensuring high quality production. The ingredients of concrete blocks are cement, sand, aggregates and water. Concrete blocks can be manufactured in varied shapes and sizes by manual or mechanical ways. The common sizes of precast concrete blocks are 400 mm, 500 mm or 600 mm in length, 200 mm or 100 mm in thickness (height) and 100 mm, 200 mm or 300 mm in width. Concrete blocks are widely used in construction as these have a better quality due to good workmanship in manufacturing units.

3.4.1 Advantages of Precast Concrete Blocks

1. Precast concrete blocks can be produced in desired shape and size as per the requirement.
2. The precast concrete blocks have high compressive strength. The sand content in the blocks is increased whereas the coarse aggregates and water content is decreased than ordinary, concrete. This resulting high compressive strength of blocks.
3. The concrete blocks are superior in durability properties like better resistance to rain, fire and environmental attack.
4. Precast concrete blocks are made in a factory and hence the overall construction speed is faster using these blocks

5. Precast concrete blocks may be made hollow to reduce its self weight, accommodate reinforcement and to improve thermal insulation.

3.4.2 Types of Precast Concrete Blocks

Precast concrete blocks are broadly classified in three categories:

(i) **Solid Blocks:** A typical precast concrete blocks is solid in nature. These blocks are used as masonry units to bear loads. The solid blocks are laid over one another and are connected to each other with mortar. The process is very similar to laying of bricks or stone masonry. However, the self weight of a solid concrete block is too high i.e., of the order of 17 – 20kg/unit which results in transporting and handling problem. Another limitation with solid concrete blocks is that these blocks have no scope of placing reinforcement bars.

(ii) **Hollow Blocks:** Hollow concrete blocks have low self weight, of the order of 10 – 12kg/unit. These are produced by keeping hollow space between the surfaces of the block so that reinforcement bars can be passed through these blocks. Hollow blocks, due to air space in between the surfaces, have good thermal and sound insulation. The space within the hollow blocks may be filled with concrete. Hollow concrete blocks are used wherever large bearing surfaces are needed.

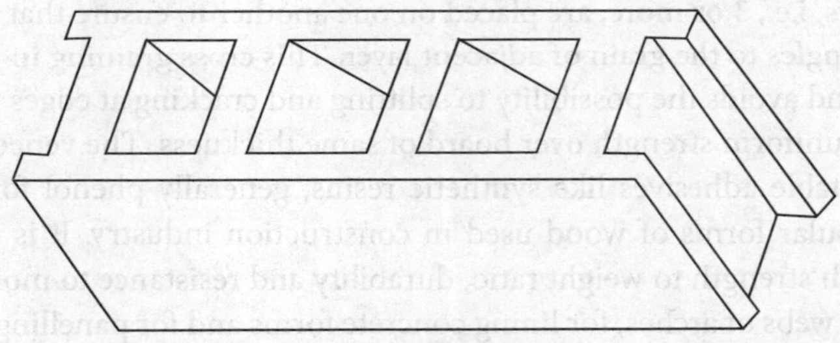

Standard Block

Figure 3.4: A standard hollow block

(iii) **Pavement Blocks:** Precast concrete solid blocks which are used for paving the surface of pavements/roads or streets are called pavement blocks. These units come in varied shapes, sizes, colour and thickness depending on the usage and aesthetic appeal required. Pavement blocks are laid over a bed of sand and are jointed to each other through interlocking without any mortar or binding material. Paver blocks are easy to lay and results in rapid constructions of pavements. Paver blocks are easy to lay/maintain and repair. These blocks are preferred for parking, footpaths or low traffic carrying pavements.

3.4.3 Uses of Precast Concrete Blocks

The precast concrete blocks have following common applications:

(a) Solid concrete blocks are used as buildings construction units in making of walls, lintels, columns and other load carrying elements.

(b) Hollow concrete blocks are also used in making of walls, partition walls, fireplace lining, tunnel lining and water tanks etc.

(c) Pavement blocks are used as construction unit in roads, parking lots and streets.

3.5 PLYWOOD AND ALLIED PRODUCTS

Timber, a natural construction material, when prepared and treated in factories to meet the requirements of desired shape, appearance, size and strength, results in some allied products like plywood, particle board, veneer, laminated board etc. These products have specific uses in building construction industry.

3.5.1 Veneers and its Uses

Veneer is a very thin slice of wood. These are produced by cutting tree logs of superior quality. Some of the timbers suitable for making veneer include teak, sheesham, rose wood, walnut etc. The thickness of veneer varies from 0.3 to 0.8 mm. Veneers are to be joined to produce plywood, boards etc. veneers are available in a verity of colour, shades and designs and hence are popular construction material for interior decorative works. Also, these have the advantage of sound insulation, durable and resistance to wrapping. Veneers can be painted or polished for achieving the desired look.

3.5.2 Plywood

Plywood is an engineered wood made by pasting thin sheets of wood i.e., veneers into panels. Odd number of veneer sheets, i.e., 3 or more, are placed on one another to ensure that the grain of each layer is maintained at right angles to the grain of adjacent layer. This cross graining in plywood increases the flexibility of plywood and avoids the possibility to splitting and cracking at edges when nailed. Plywood has advantage of more uniform strength over board of same thickness. The veneer sheets are cemented with each other by suitable adhesives like synthetic resins, generally phenol formaldehyde. Plywood is one of the most popular forms of wood used in construction industry. It is used as wall and roof sheathing due to its high strength to weight ratio, durability and resistance to moisture. Plywood is also used in sub flooring, in webs of arches, for lining concrete forms and for panelling in interior decorative works. Besides, plywood can make furniture as it is inexpensive and easy to work with. It is light in weight and easy to move.

3.5.3 Particle Board

These are prepared by processing pieces of wood or other organic fibres and pressing them to get board of suitable thickness. It is also called pressed board or fibre board. Another type of board is block board which is prepared from strips of wood glued to form a slab which is further placed between two or more veneers maintaining cross graining between adjacent layers and then pressed to form block board. Boards are used for wide range of purpose in construction such as for wall panelling, partition walls, flooring and suspended ceiling and also in ornamental work, furniture and table tops etc.

3.5.4 Laminated Board

Laminated board is prepared by joining the ends of short boards by strong adhesives. These are popular because obtaining large size or sections of timber is difficult and hence laminated board are made using small piece of board. Laminated board is a versatile wood based product which can be used for structural applications. Its advantages include desired shape and size, homogeneity, high strength and low cost with minimum number of defects. Laminates may be bent during gluing process in curves and such laminates are used in trusses, arches and other architectural places. Further, laminated boards can be used to enhance the look of surface of walls and furniture.

3.6 GLASS

Glass is one of the most versatile and ancient material in construction industry. Its use was earlier limited to window panes but now a day, it is also used as a structural building material. Glass is a mixture of number of metallic silicates. The use of alkali metals is popular in making of glass. A variety of glass types can be produced depending on the desired usage. Glass can be made as soft as cotton, as strong as steel, as light as cork and many more.

3.6.1 Properties of Glass

Glass has major properties as below:

(i) Glass is typically a hard, brittle and transparent material.

(ii) It has good resistance to water, air and chemicals.

(iii) It has no sharp melting point.

(iv) Glass has excellent sound and electric insulation.

(v) Fusion of glass is possible.

(vi) It can be made in different colour, Shapes, Sizes and textures.

(vii) Glass absorbs or transmits light depending upon its type for a particular use.

3.6.2 Types of Glass

Glass can be prepared in a big variety of types. Some of the commonly used types of glass are:

- **Soda Lime Glass:** This glass is commonly known as float glass. It is manufactured from sodium silicate and calcium silicate. Soda lime glass is very clear and flat and hence causes glare. Float glass is available in varied thickness i.e., of the order of 2 mm to 20 mm. This glass is used in front panels of shop or public places to give clear visibility inside.
- **Shatter Proof Glass:** This type of glass is manufactured by adding some type of plastic polyvinyl butyral. This glass does not have any sharp edges pieces when it breaks. Such glass has great utility for windows floors and skylights etc.
- **Laminated Glass:** As its name suggests, laminated glass is made by a combination of various layers (laminates) of ordinary thin glass. Laminated glass has more weight and thickness. Laminated glass has the advantage of sound proof and UV protection. This glass is used for aquariums etc.
- **Toughened Glass:** Toughened glass is one of the most durable types of glass which can resist vibrations and does not break easily. It is available in various thicknesses and is used for fire–resistant door, mobile screen protectors and table top etc.
- **Lead Glass:** Lead glass is manufactured by adding lead oxide in place of calcium oxide in ordinary glass. Generally, the content of lead oxide is kept less than 30%. Earlier, lead glass was used to make glassware like drinking glass, jewellery, ornament etc. Lead glass has crystal clear properties which makes it popular for decorative purposes. But, now-a-days lead glass glassware is not preferred as there is potential health risks involved due to use of lead oxide. Now, lead glass is made with reduced lead content or lead free glass by adding zinc oxide or potassium oxide to

substitute lead oxide. This glass is used for making healthcare and scientific shielding for workers working under radiation exposure industry.

- **Borosilicate Glass:** Borosilicate glass consists of substantial amount of silica (SiO_2) and Boron oxide (B_2O_3). Borosilicate glass is more durable and resistant to chemical and temperature changes and hence is used to make apparatus for chemistry labs, Kitchens and premium drinking glasses. This glass can be easily milled, drilled, grounded and made toughened. However, the use of borosilicate glass is not popular in construction become it is known weaker against mechanical forces and breaks easily.

3.7 FERROUS AND NON FERROUS

All the metals or alloys which contain iron as their main constituent are called ferrous metals. For example– Cast iron, wrought iron, steel etc. However, those metals which do not contain iron as the main constituent include– Aluminium, brass, lead, copper etc.

3.7.1 Ferrous Metals and Their Uses

- **Cast Iron:** Cast iron is manufactured by refining pig iron to bring carbon content in the range of 1.7 – 4.5%. Cast iron has clear crystalline structure. Cast iron is brittle and hard. It is strong in compression and weak in tension. Cast iron does not absorb vibration and shocks. Other limitations of cast iron are that it cannot be welded, rolled, punched or riveted. But, the advantage of cast iron is that it has excellent resistance to corrosion and it rusts slowly. Cast iron is used for making columns, bases and caps of columns, brackets, sewage pipes, sanitary fitting, manhole covers, rain water pipes and spiral stair cases etc.
- **Wrought Iron:** Wrought iron is the purest form of iron with low carbon content of the order of 0.15% or less. The process of removing carbon, manganese, silicon and sulphur from pig-iron is called 'Puddling'. Wrought iron is ductile, tough and malleable. It is more prone to rusting than cast iron. Wrought iron has good compressive as well as tensile strength. Wrought iron is used to make nails, bolts, nuts, chains, sheets, ties, handrails, plates, trusses and pipes etc.
- **Steel:** Steel is an alloy of iron and carbon as carbide of iron. It has a carbon percentage limited to 1.5%. Lesser the carbon content in steel, its properties are closer to wrought iron and more the carbon, steel resembles cast iron in properties. Steel is highly elastic, ductile, weld-able, malleable and forgeable. Steel has wide usage in building construction industry. It is used as a structural material in the form of reinforcing bars, trusses, frames and beams. As non structural and non load carrying members, steel is used to make grills, stairs, windows and doors, frames etc. Steel is also used to make pipes, sewer fittings, corrugated sheets, ducts and tanks etc.

3.7.2 Non Ferrous Metals and Their Uses

- **Aluminium:** It is a white/silvery shining metal obtained from 'Bauxite' ore. Aluminium, in its pure form, is very soft and ductile. Aluminium is mixed with other metals like copper, magnesium, manganese etc, as an alloy, to increase its hardness and tensile strength. Aluminium is light in weight, durability, highly resistant to corrosion and good conductor of heat and electricity. It is used as reinforcement material in tall building, corrugated sheets for roofing, window and door frames, posts, balustrades etc.

- **Copper:** Copper is obtained by smelting 'Copper Pyrites' – an ore of copper. Copper has a lustrous red colour and has many properties which make it a useful material for engineering purposes. It is light in weight, tough, strong, ductile, forgeable, rollable and malleable. It is a good conductor of heat and electricity. Copper is widely used as electric wire, cables, tubing for water supply and gas, sheets for roofing, in DPC as sheets and also for electroplating.
- **Brass:** Brass is an alloy of copper and zinc. The most common commercial brass has 65% copper and 35% zinc. Brass is bright yellow metal which resists corrosion. It has almost all the advantages of copper like ductility, malleability, casting and forgeable etc. Brass is used as fillings to doors and windows, stop cocks and valve in water fitting, household utensils and number/name plates etc.

UNIT SUMMARY

- Bricks
  - Standard Size – 190 mm × 90 mm × 90 mm
  - Major Constituent – Clay and Silica
  - Minor Constituent – Lime, Iron Oxide, Magnesium
  - New Type – Fly ash bricks containing 50 – 60% fly ash
- Characteristic of good brick
  - Sharp Edges
  - Uniform Shape and Size
  - Soundproof
  - Good Strength
  - Fire Resistance
  - Durable
- Classification of Bricks
  - First Class Bricks – Best Quality
  - Second Class Bricks – Reasonably Good Quality
  - Third Class Bricks – Not Preferred
  - Over-burnt Bricks – Not Recommended
- Manufacturing of Bricks – Steps involved are:
  - Preparation of Clay
  - Moulding of Clay
  - Drying
  - Burning
- Aerated Concrete Blocks
  - Make up of Fly ash, Lime, Cement and Gypsum
  - Advantages: Light in Weight, Easy to Cut, Better Sound and Thermal Insulation, Eco-friendly and sustainable
- Types of Flooring Tiles
  - Ceramic Tiles: made up of clay, burnt at high temperature, commonly used for flooring.

    - Porcelain Tiles: made up of heavy and denser clay, Superior and costlier to ceramic tiles, preferred under heavy traffic.
    - Cement Floor Tiles/Mosaic Tile: made up of cement, available in decorative colour and pattern, are water tight and hard.
    - Quarry Tiles: made up of clay, highly resistant to chemical attack, used in laboratories, industries and tanneries for flooring.
- Cement
    - Major Components: Lime and Silica
    - Minor Component: Alumina, Iron oxide, Magnesium oxide and Sulphur oxide
    - Manufacturing: Wet Process and Dry Process
- Types of Cement
    - Ordinary Portland Cement (OPC): Most commonly used, available in three grades – OPC_{33}, OPC_{43}, OPC_{53}
    - Portland Pozzolana Cement (PPC): obtained by grinding of OPC and pozzolana, has low early strength but comparable ultimate strength, better resistance to sulphate and chemical attack.
    - Rapid Hardening Cement (RHC): gains strength faster, 3 days strength of RHC is equal to 7 days strength of OPC, finer than OPC.
    - Quick Setting Cement: sets faster than OPC but gains strength at the same rate.
    - Low Heat Cement: has Low heat of hydration, suitable for mass concreting – dams, bridges etc.
    - Sulphate Resisting Cement: has better resistance to alkali attack, suitable for under soil construction.
    - Blast Furnace Slag Cement: obtained by grinding OPC and blast furnace slag, gains strength slowly, has low heat of hydration, better resistance to sulphate attack.
    - White Cement: used for decorative works.
- Field Test for Cement
    - Colour
    - Lump-free
    - Feel
    - Smoothness
    - Sink Test
    - Glass Plate Test
- Pre-Cast Concrete Blocks
    - Types: Solid Blocks, Hollow Blocks, Pavement Blocks
    - Advantages: High Compressive Strength, Durable, Good Resistance to rain and fire, can be made in desired shape and size.
- Allied Product of Timber
    - Veneer: Thin slice of wood, good strength, durable and resistance to wrapping, good sound, insulation property.
    - Plywood: Obtained by pasting multiple layers of veneers, available in varied thickness, strong, durable and versatile inexpensive material.
    - Particle Board: Obtained by pressing wooden pieces under pressure. Light in weight and highly useful for partition walls, furniture etc.

- Laminated Board: Obtained by pasting small board pieces by strong adhesives, used in arches, trusses etc.

- Glass: Hard, brittle and transparent material consisting of metallic silicates
  - Advantages: Transparent, no sharp melting point, good resistance to water, air and chemicals, excellent sound and electric insulation.
  - Types of glass: Soda Lime Glass, Laminated Glass, Lead Glass, Toughened Glass, Borosilicate Glass.
- Ferrous Metal: Contains Iron
  - Cast Iron: Carbon between 1.7 – 4.5%, Good resistance to corrosion, used for sewage pipes, water supply and sanitary fitting, Manhole covers etc.
  - Wrought Iron: Carbon content less than 0.15%, ductile and malleable, prone to rusting, used for making nails, nuts, bolts etc.
  - Steel: Alloy of iron and carbon, carbon content less than 1.5%, ductile, weldable, forgeable, malleable, widely used in construction sheets, ducts, tanks etc.
- Non-Ferrous Metals: Does not contain iron as main constituent
  - Aluminium Silvery white metal obtained from Bauxite ore.
  - Copper: Lustrous red colour metal obtained from copper pyrite ore.
  - Brass: Bright yellow colour, alloy of copper and zinc.

EXERCISES

Multiple Choice Questions

1. The primary purpose of 'frog' in a brick is to
   (a) Reduce the weight of the brick
   (b) Improve insulation
   (c) Form a keyed joint between the mortar and the brick
   (d) Inscribe the manufacturing firms initials
2. The weight of a standard brick should be
   (a) 1.5 kg (b) 2.0 kg (c) 2.5 kg (d) 3.0 kg
3. The rate of hydration of cement depends upon its
   (a) Compressive Strength (b) Soundness
   (c) Fineness (d) Slump
4. Gypsum is added in the manufacture of cement in order to
   (a) Decrease the burning temperature (b) Decrease the grinding time
   (c) Decrease the setting time (d) Increase the setting time
5. For the manufacture of plywood, veneers are placed such that grains of advancement veneers are
   (a) Parallel (b) Inclined at 45°
   (c) Inclined at 30° (d) At right angles

6. A product obtained by assembling veneers with adhesives is called
   (a) Plank (b) Batten (c) Board (d) Plywood
7. The process by which wrought iron is obtained from pig iron is known as
   (a) Annealing (b) Puddling (c) Shingling (d) Forging
8. Brass is an alloy of
   (a) Zinc and Lead (b) Zinc and Nickel
   (c) Silver and Tin (d) Zinc and Copper
9. Rapid hardening cement has
   (a) Faster rate of strength gain (b) Faster setting
   (c) Low heat of hydration (d) Better resistance to alkali
10. Which of the following has better acoustics
    (a) Bricks (b) Aerated Concrete Blocks
    (c) Tiles (d) Plywood
11. Which types of tiles have good resistance to acid or alkali attack
    (a) Ceramic Tiles (b) Porcelain Tiles (c) Quarry Tiles (d) Cement Tiles
12. Clinker is a term associated with
    (a) Timber (b) Cement (c) Tiles (d) Plywood
13. Ordinary cement has which component content less than 1%
    (a) Lime (b) Silica (c) Alumina (d) Iron Oxide
14. Maximum ductility is found in
    (a) Glaze (b) Plain Steel (c) Fibre Glass (d) Cast Iron
15. The maximum carbon content in structural steel should be
    (a) 2% (b) 1.3% (c) 3% (d) 4%
16. The property of a material due to which it can be drawn into thin wires is known as
    (a) Hardness (b) Ductility (c) Toughness (d) Malleability
17. The crudest form of iron is
    (a) Pig Iron (b) Cast Iron (c) Wrought Iron (d) Mild Iron
18. The biggest disadvantage of glass is
    (a) Strength (b) Durability (c) Brittle (d) Hardness
19. The weight of a hollow concrete block is of the order of
    (a) 10 – 12 kg (b) 17 – 20 kg (c) 25 – 30 kg (d) 5 – 7 kg
20. When hand is immersed into a bag of cement, it should feel
    (a) Hot (b) Warm
    (c) Cool (d) None of the above

ANSWERS																			
1.	(c)	2.	(d)	3.	(c)	4.	(d)	5.	(d)	6.	(d)	7.	(b)	8.	(d)	9.	(a)	10.	(b)
11.	(c)	12.	(b)	13.	(d)	14.	(d)	15.	(b)	16.	(b)	17.	(a)	18.	(c)	19.	(a)	20.	(c)

Short Answer Type Questions

1. What are the major constituents of brick earth? State the function of each constituent.
2. Compare bricks with aerated concrete blocks describing their advantages and limitations.
3. What are the characteristics of first class bricks?
4. Why are tiles preferred in making floors? State the advantages.
5. How do you test the quality of cement in field?
6. Draw a flow chart describing manufacturing of cement by wet process.
7. Write short notes on:
   (a) Rapid Hardening Cement (b) Blast Furnace Slag Cement.
8. What do you mean by pre-cast concrete blocks? State their types and uses.
9. What are fly ash bricks? Where are these used?
10. State the composition of ordinary Portland cement.

Long Answer Type Questions

11. Explain the manufacturing of burnt clay bricks in details.
12. What are various products allied with plywood? Explain about each in brief and state their uses.
13. What are the types of glass? Which type of glass is most popular in construction industry?
14. What Non-Ferrous Metals are used as building material? State their uses.
15. Compare various types of ferrous metals used in construction.

PRACTICAL

Experiment-1

Aim: Select first class, second class and third-class bricks from the stake of bricks and prepare report on the basis of its properties

Materials Required: Sample bricks, say 20 in number

Theory: Bricks are classified into three major categories– first class bricks, second class bricks and third class brick. This classification is based on the quality and extent of burning of bricks in kiln. These bricks have distinct characteristics on the basis of which these can easily distinguish from each other. A comparison chart of these three categories of bricks is as follow:

Property	1st Class Bricks	2nd Class Bricks	3rd Class Bricks	Over burnt Bricks
Edges and Corner	Sharp	Rounded	Irregular	Fused
Surface	Smooth	Rough	Very Rough	Coarser
Colour	Crimson Red	Yellowish Red	Light Red	Dark/Blackish Red
Sound	Metallic Sound	Metallic Sound	Dull Sound	Blunt Sound
Burning Status	Thoroughly brunt	Uniformly burnt	Under burnt	Over burnt

Property	1st Class Bricks	2nd Class Bricks	3rd Class Bricks	Over burnt Bricks
Tolerance Limit in Crushing Strength	3%	8%	10%	–
Efflorescence	Nil	Slight	Considerable	Too high
Water Absorption	<10%	<15%	15 – 20%	>20%

Procedure:

1. Take the sample bricks one by one.
2. Observe each brick very carefully for its colour, shape, edges, size, texture, burning status and efflorescence.
3. Record the above parameters for each sample bricks.
4. Strike two bricks against each other and observe the sound.
5. Classify the given bricks according to its characteristics.

Observations:

Sample No.	Characteristics						
	Colour	Edges (Sharp/ Rounded/Fused)	Surface (Smooth/ Rough)	Sound (Metallic/ Dull)	Burning (Uniform/ Uneven)	Efflorescence (Slight/ Considerable)	Remarks (Class 1st/2nd/3rd)
01							
02							
03							
04							
05							
06							
07							
08							
09							
10							
11							
12							
13							
14							
15							
16							
17							
18							
19							
20							

Conclusion: Bricks can be classified into first/second/third class by visual inspection.

Precautions:

1. The sample bricks should be free from any pasting/covering material on their surface or should be cleaned before making observations.
2. The sample bricks should be observed very carefully.

Experiment-2

Aim: Measure dimensions of 10 bricks and find average dimension and weight. Perform field tests: dropping, striking and scratching by nail and correlate the results obtained.

Materials Required: 10 sample bricks, Scale, Weighing Balance with an accuracy of 100 gm.

Theory: The quality of bricks can be accessed in the field by various tests like measuring dimensions, weight, striking (Sound test), dropping test or scratching with nail test. These tests can give fair judgement on the quality of brick in terms of its strength and durability. The bricks can be correlated to various classes i.e., 1st/2nd/3rd class by the results of these tests. The tests are commonly adopted in field.

Procedure:

1. Take a sample of 10 bricks for observations.
2. Measure all the three dimension of each brick carefully and record in observation table.
3. Weigh each brick in weighing balance up to an accuracy of nearest 100 gm.
4. Strike two bricks against each other and record the sound produced.
5. Try to scratch each brick with your finger nail and record the impressions, if any.
6. Drop each brick from a height of about 1 meter and record its breaking status.

Observations:

Sample No.	Dimension (in mm) (L) (B) (T)	Weight (in grams)	Sound generate (metallic/dull)	Nail impression (No/Yes)	Dropping breakage (broken/not broken)	Quality of bricks (Good/Average/ Poor)
01						
02						
03						
04						
05						
06						
07						
08						
09						
10						

Conclusion: A good brick should have following characteristic.

Dimension	Weight	Sound Generated	Nail Impression	Dropping Breakage
190 mm × 90 mm × 90 mm	3 – 4 Kg	Clear Metallic ringing sound	No Impression	Should not break

Bricks can be correlated to their quality by these field tests.

Precautions:

1. The sample bricks should be free from any covering/pasting on their surface.
2. The bricks should be cleaned before taking observations.
3. The observations should be recorded very carefully.

Experiment-3

Aim: Identify different types of flooring tiles such as vitrified tiles, ceramic tiles, glazed tiles, mosaic tiles, anti-skid tiles, checkered tiles, paving blocks and prepare report about the specifications.

Materials Required: Sample Tiles, Say 20 in number, Magnifying Glass

Theory: Flooring tiles are available in a variety of types. Selecting suitable tiles for a project/specified purpose is very important. Tiles are not easy to identify as most of flooring tiles are made up of clay and other materials and are kiln burnt. However, a few simple observations on colour, smoothness at joints, glaze and resistance to stain can give a fair idea about the type and quality of tile. The standard observations for varied types of tiles are as follows:

Types of Tile	Surface Finish	Glaze	Colour of Side/Base	Stain Removal	Faces/Edges
1. Vitrified/ Porcelein Tiles	Fine grained/ Smooth	Same colour throughout thickness	Any Colour but not white/ Tan	Easy	Sharp Edges, Uniform face
2. Ceramic Tile	Bumpy	White/tan base	White/tan/red	Difficult	Rounded edges, non-uniform colour face
3. Glazed Tile	Smooth	Tan/Red base	White/tan/red	Easy	Sharp edges
4. Mosaic Tile	Coarse Finish	White base	–	Moderate	Rounded edges
5. Anti-skid	Rough	–	–	Tough	Non-uniform faces
6. Paving Blocks	Coarse rough surface	Uniform colour throughout	–	Tough	Non sharp edges

Procedure:

1. Take the sample tiles one by one.
2. Clean the surface of tiles with a wet cloth.
3. Closely observe the surface finish of the tiles by running your fingers over the top surface of tiles, for smooth, bumpy or coarse finishing and report carefully for each tile.
4. Observe closely at the glaze of each tile. If the tile is chipped, its white or tan (coloured) base can be observed. Record the observations.
5. Inspect the faces and edges of each tiles and record for sharpness and uniformity.

6. Examine the sides of each tile for a white, tan or reddish colour base.
7. Examine the tiles for any kind of stains on it. A few drops of some colouring dye can be dropped on tile to see its absorption. Try to wipe off the dye stains after few minute. Record the observations.

Observations:

Sample No.	Surface finish (Smooth/ Bumpy/ coarse)	Glaze	Colour of sides/base	Stain removal	Faces/Edges	Remarks
01						
02						
03						
04						
05						
06						
07						
08						
09						
10						

Conclusion: The tests performed on tiles are based on visual inspection and hence are for approximate judgement and are not conclusive in nature.

Precautions:

1. Always clean the surface of tiles before observations.
2. Observe the tile very carefully.

Experiment-4

Aim: Identify the type of glasses from the given samples.

Materials Required: Sample pieces of glass (Say 10), magnifying glass, cutting tool.

Theory: Glass is a mixture of a number of metallic silicates. The varied types of glass are available in market. The glass can be identified for its type by performing some simple tests in field or laboratory. These tests involve inspection of colour, thickness, clarity, hardness and light transmission etc. The general observations of a variety of glass are as follows:

Sr. No.	Glass Type	Edges	Imperfections	View under polarized glass	Surface after cutting
01.	Soda Lime/ Float glass	Ridge edges	No scratches	clear	Smooth cut
02.	Shatter proof glass	Not sharp	Few scratches	Light coloured	Smooth cut
03.	Laminate Glass	Rough edges	scratches visible under magnifying glass	–	Slight uneven line

Sr. No.	Glass Type	Edges	Imperfections	View under polarized glass	Surface after cutting
04.	Toughened/ Tempered Glass	Perfectly smooth	Visible bending/dimples	Dark shady spots	Uneven flaky line
05.	Lead/crystal glass	Sharp edges	Uneven surface/ dimples	clear	Uneven flaky line
06.	Borosilicate glass	Sharp edges	No scratches	–	Straight line

Procedure:

1. Take the sample pieces of glass one by one.
2. Clean the surface of each glass piece with a wet cloth.
3. Examine the edges of the glass for its smoothness or ridges or sharpness and record the observations.
4. Look for any stamp or bug or label etched or sand blasted on the glass. These labels sometime indicate the manufacture's name, safety standards and type of glass.
5. Inspect the glass pieces for any kind of imperfections due to bending/wrapping or over working on it and record the same.
6. View the surface of glass in sunlight under a polarized lens or goggles and record the colour.
7. Try to score/mark a line on the surface of glass using a window glass cutting tool and record the shape/smoothness of the line.

Observations:

Sample No.	Edges (Ridge/ rough/ smooth)	Imperfection (Wrapping/bending/ dimple/no imperfection)	Under polarized lens (Colour)	Surface after cutting (smooth/flaky)	Remarks (Type of glass)
Sample 01					
Sample 02					
Sample 03					
Sample 04					
Sample 05					
Sample 06					
Sample 07					
Sample 08					
Sample 09					
Sample 10					

Conclusion: The tests performed on glass are based on visual inspection and hence are for approximate judgement and are not conclusive in nature.

Precautions:

1. Always clean the surface of glass before observations.
2. Observe the glass very carefully.

KNOW MORE

The use of artificial construction materials is of great significance in all types of infrastructural projects. The choice of material typically depends upon the climate and region of the proposed project. Besides, the extent of availability of the natural building materials also affects the dependency on artificial building materials. Today, for the ease of construction, speed of the construction and for the desire of comfort levels, a; large number of artificial construction materials are used in each and every project. The fact is that the se of artificial construction materials has surpassed the volume of use of natural materials. At many places, it is observed that the aesthetic appeal, decorative looks, the kind of finishes required and the tourism attraction play an important role in deciding the selection of construction material. But, one must keep the following parameters in mind for selecting the appropriate material- safety first i.e. sufficient strength, durability second i.e., long life span, economy third i.e. cost- effective and comfort last i.e. easy to use. The order of these parameters should not be altered.

REFERENCES AND SUGGESTED READINGS

- Rajput, R.K, Engineering Materials, S. Chand and Co., New Delhi.
- Sood H., Laboratory Manual on Testing of Engineering Materials, New Age Publishers, New Delhi.
- Sharma C. P., Engineering Materials, PHI Learning, New Delhi.

4

UNIT SPECIFICS

This unit specifies the following aspects on special construction materials:

- Knowing some special construction materials.
- Understanding the suitability of special construction materials for specific purposes.
- Knowing various types of fibres used in the construction.
- Understanding the uses of various types of fibres.
- Knowing geo polymer cement and understanding its properties and uses.

Besides, the stated aspects, an experiment related to termite proofing of a damaged timber sample has been explained in this unit. Termite attack on timber is a common problem and a lot of measures have been tried since ancient times to make timber termite proof. The information on present day treatment techniques to protect timber from termite attack will help young budding civil engineers to tackle this common recurring problem effectively and efficiently especially in an apartment constructions. This chapter also discusses the various types of fibres and their uses in the construction industry and thus, helps students to select a suitable fibre for a specific use.

Objective type multiple choice questions have been placed at the end of the unit to assess the level of understanding by the students on special construction materials. Besides, short answer type and long answer type questions, based on bloom's taxonomy have been given in the last section of this unit.

Know more section and suggested readings through QR codes have also been designed to create an inquisitive approach in young minds on special construction materials. A practical on application of relevant termite proof chemical to safeguard a damaged timber sample has also been explained at the end of this unit. The hands-on-practice by the students in the laboratory will definitely give them more confidence and expertise in handling one of the most common problems of termite attack.

RATIONALE

This unit, basically, deals with three subcategories of special construction materials. The first one is related to special materials used for water proofing of existing structures, termite proof treatment on existing timber structures and use of thermal and sound insulating materials. The next section in the unit is related to the use of fibres in construction. Various types of fibres and their suitability have been covered very precisely in this section. The third subcategory in the unit is related to geo-polymer cement, its properties and uses. All these special materials are adopted for a special purpose either as an alternative or as an additive to popular construction materials. A rationale on the use of special construction materials can be sought by civil engineering students after understanding this chapter for solving the prevalent challenges of dampness, termite attack, shrinkage, and other related environmental issues.

PRE-REQUISITE

Knowledge on natural and artificial construction materials (i.e., Unit-II and Unit-III)

UNIT OUTCOMES

Following is the list of outcomes from unit on special Construction Materials:

U4-O1: Understand the need for water proofing, termite proof, thermal and sound insulation.

U4-O2: Define and know various special materials for above mentioned requirements.

U4-O3: Know various types of fibres and understand their usage in construction.

U4-O4: Understand properties and uses of geo-polymer cement.

U4-O5: Apply the knowledge gained to achieve water proofing and insulation in structures.

Unit-4 Outcomes	EXPECTED MAPPING WITH COURSE OUTCOMES (1- Weak Correlation; 2- Medium Correlation; 3- Strong Correlation)				
	CO-1	CO-2	CO-3	CO-4	CO-5
U4-O1	1	2	2	3	1
U4-O2	3	2	3	3	2
U4-O3	3	2	2	3	2
U4-O4	3	2	3	3	2
U4-O5	1	2	2	3	1

4.1 INTRODUCTION TO SPECIAL CONSTRUCTION MATERIALS

The construction materials which are used for a specific purpose as an alternative or additive to commonly used building materials are called special construction materials. Some of the common challenges faced by structures are dampness, attack on its timber elements by termite or the requirement of making the structure thermal resistant and sound resistant. The use of special construction materials may be incorporated while building a new structure or can be treated as an activity for repair, restoration or strengthening the existing structure. The performance and life of any natural or artificial material depend on its usage and its exposure to attacking agencies like moisture, insect attack or environmental attack. Special construction materials can be used to control or avoid these consequential effects on the building.

4.1.1 Water Proofing Material

The building surfaces that are exposed to water/moisture have a tendency to get penetrated by water which causes dampness in the structure. The penetrated water not only gives a poor look but also may result in leaking from the surface. Water proofing is a technique to form an impervious layer over the surface of various elements like foundation, roof, walls, columns and other structural members. This impervious layer prevents the entry or penetration of water or moisture in the structure. This impervious membrane can be internally or externally developed. Internal membranes are created by use of water proofing admixtures whereas external membranes can be formed by products like paints, grouts, sealants etc.

(i) **Admixtures:** Admixtures helps in water proofing of concrete by two ways. Firstly, some of the water reducing admixtures reduce the water content in concrete and make the concrete dense, compact, and more durable and water proof. Secondly, some of the admixtures result in creating water tight coating on the surface of the concrete, making it water proof.

(ii) **Impregnation:** Impregnation is a technique in which some solution is penetrated into pores of the structure, resulting in a water tight layer. This technique is useful for water proofing of both old as well as new structures. There are three types of impregnations. These are hydrophobic phase, partial filling of pores and full filling of pores. For hydrophobic type, saline, quartz carbine solutions are used. Silicone and sodium silicate solution are used in partial filling techniques. Low viscosity epoxy and methacrylate solutions are used for filling the pores completely.

(iii) **Surface Film:** Asphalt, epoxy mortar concrete or some elastomeric membranes are the materials used to form a water proof film on existing concrete surfaces.

(iv) **Sealants:** Joints act as a weak zone and cause water penetration. Various joints like column-beam joint, column-slab joint, beam-beam joint or slab-slab joint etc. are sealed properly by suitable sealants.

4.1.2 Termite Proofing

Termite is also known as white ants. Termite attack is one of the major problems faced in any structure, especially on its wood work. The wood work of a building includes doors, windows, furniture, cupboards, finishing or sometimes decorative interior work. Termite, besides wood, may also damage some non-cellulose materials like plastic, paper, rubber or leather. It is very important to control termite attack as the damage caused by termite is huge. The termite damage in wooden works is more and faster as cellulose present in wood acts as nutrients or food to termite/white ants.

4.1.2.1 Types of Termites

On the basis of their habitat, termites are classified into two groups:

(i) **Subterranean Termite:** These are also known as ground-nesting termite. Such termites build their nest in the soil having adequate moisture. This type of termite is found in almost all regions of India. These termites make a route from soil to building through its foundation. Subterranean termite eats all cellulose containing materials i.e., wood, paper, cardboard and even cloths.

(ii) **Non-Subterranean Termite:** These are also known as dry-wood termites. Such termites live on wood and have no contact with soil. These live in small colonies and attack wood work directly. The dry-wood termite attack may lead to holes in the wood or may damage the whole wooden work by reducing it to powdery form.

4.1.2.2 Principles of Termite Proofing

The general principles for making a building termite proof are as follows:

- Suitable termite proofing treatment should be given to the building either pre-construction or during construction.
- Proper site drainage should be ensured to drain away rain water.
- The gap between foundation and plinth should be filled with termite proof chemicals or materials.
- Joint filers or metal strips used to seal floor joints should be termite proof or treated with termite proof materials.
- No gap between any part of the building and untreated soil should be ensured to avoid termite route into the structure.

4.1.2.3 Stages of Termite Proofing

Termite proofing or anti termite treatment in buildings is carried out in two broad stages:

(a) Pre-Construction Treatment. (b) Post Construction Treatment.

4.1.2.3.1 Pre-Construction Treatment

Pre-construction anti termite treatment is considered the most effective way to prevent termite attack in the building. In this method, the soil under and around the foundation of building is treated with chemicals. These chemicals form a barrier between the ground and the brick work of the foundation and prevent the invasion of termite into the building. The various steps involved in this type of treatment are as follows:

(i) **Site Preparation:** The site for construction is removed from any logs, roots, stumps or waste wood etc. If any sign of termite mounds (groups) is detected within the plinth area of the building, these are destroyed by spraying suitable insecticide chemicals.

(ii) **Soil Preparation:** The bottom and sides of the trenches excavated for laying foundation are treated with chemicals like concentrates of chloropyrifos. The refilled earth near the walls and columns should also be treated with anti-termite chemicals.

(iii) **Structure Barrier:** The structural barriers should be created at the plinth level to prevent termite entry into the building through walls. The barrier can be made either by rich concrete layer or by sheets of non-corrodible metals like copper or galvanized iron. Besides plinth level, expansion joints and periphery of pipes and conduits should be treated with anti-termite chemicals. During installation of doors and windows the surface in contact with masonry should be coated with anti-termite paints.

4.1.2.3.2 Post Construction Treatment

This treatment is applied on existing buildings which are already under termite attack. Termites even after entering the structure, maintain a regular contact with their nest in the ground. The nest of termite in the soil needs to be destroyed by treatment of soil underneath with chemical emulsion. For treating floor and walls, holes (diameter 12 mm, spacing 300 mm) are made and these holes are soaked with anti-termite chemical solution. Later, the holes are sealed with cement mortar of 1:2. For dry wood termite, if the existing wood work is badly damaged by termites, it needs to be replaced with new timber treated with oil or kerosene based emulsion. For the undamaged wood work, a layer of suitable chemical emulsion needs to be sprayed to prevent termite attack.

4.1.2.4 Chemicals used as Anti-termite Solutions

The commonly used chemicals for termite proofing are:

- Chlorophyriphos.
- Oil or Kerosene based emulsions.
- Other chemicals like Aldrin, DDT etc may also prevent the growth of termites, but the use of these chemicals in not recommended as their usage pose significant health related hazards for human beings.

4.1.3 Thermal Insulating Materials

Thermal insulation in buildings refers to reduction of heat transfer from inside and outside of a structure. The aim of thermal insulation is to maintain desired temperature inside the building with minimum consumption of energy. The materials with good thermal resistance can block the exchange of heat between outside and inside of buildings. This ensures that the energy consumption by air conditioners or heaters/blowers for maintaining the temperatures is reduced. The heat transfer takes place in either of the following ways:

- Conduction – Transmission of heat within a material.
- Convection – Transfer of heat by other agents like air or water.
- Radiation – Transmission of heat through a wave motion.

Thermal insulating materials prevent heat transfer by convection and maintain the temperature inside the structure.

4.1.3.1 Characteristics of Good Thermal Insulating Material

Characteristics of good thermal insulating materials are:

- The insulating material should be fire proof.
- The material chosen for insulation should be resistant to insect attack.
- A good thermal insulating material should be moisture proof.
- The insulating material should be resistant to any physical change that can reduce its efficiency or effectiveness to prevent heat transfer.

4.1.3.2 Common Thermal Insulation Materials

There are a variety of materials available in the market which can be used for thermal insulation of buildings. Polystyrene foam, mineral wool, rigid panel of fibres and polyurethane foam are commonly used thermal insulators. The thermal insulation effectiveness of a material depends upon its thermal conductivity coefficient. The lower the value of thermal coefficient λ, better is the thermal insulation of the material and it controls the temperature inside the building more efficiently.

- **Mineral Wool:** It is an effective material used to ensure good thermal insulation. Mineral wool consists of two products, namely rock wool and glass wool which have very similar properties. Rock wool is obtained from basalt and the glass wool is made from quartz sand or recycled glass.

The advantages of mineral wool are that it has good insulating properties; it is very durable and has high resistance to deformation. The limitation associated with mineral wool is that it can absorb water which, in turn, can adversely affect its thermal insulating property. Skilled labour is required to install mineral wool on various elements in the buildings.

- **Polystyrene Foam:** Polystyrene foam in an expanded form of polystyrene, it involves intruding air into the pores to expand polystyrene and make a form. The advantage of polystyrene form is that, unlike mineral wool, it has very low water absorption and hence it can be successfully used to insulate those elements of building which are exposed to water such as foundation, basement walls or floors on the ground etc. Another advantage of polystyrene is that it is hard and hence can bear heavy loads. However, the limitation of polystyrene foam is that it is not resistant to chemical attacks such as those caused due to use of paints, solvents, glues or wood preservative. Moreover, this foam is sensitive to very high temperature and fire and gets damaged on exposure.
- **Polyurethane Foam (PUR):** Polyurethane foam is gaining popularity these days as an insulating material. Polyurethane foam is obtained by mixing two raw materials– Polyol and Isocyanate. This foam is sprayed directly with the help of machines onto the surface to be insulated. PUR foam performs well in terms of its efficiency, safety and functionality. The use of Polyurethane foam has the advantage of speedy and easy application. As it is used as a spray, the volume covered by the foam is much more than other materials and also this foam can reach even the difficult, non-accessible and small gap surfaces. PUR is used in almost all parts of a building for insulation purposes. These include foundation, walls, slabs, halls, basements and even heating pipes, cooling chambers, tanks etc. in industries.
- **Cellulose Fibres:** Cellulose fibres are very similar to mineral wool in characteristics but these have a good ability to absorb and release water from the surroundings. Fibres can be applied by wet or dry application. In dry method, loose shredded cellulose fibres are blown into the pre-conditioned spaces in walls and ceiling to act as insulation. In wet method, the fibres are made wet with water and adhesive and this paste is poured into spaces prepared in walls and roofs.
- **Wood Wool:** The fibres of wood are bonded with cement to form stable sheets which are then used as thermal insulation material. The advantages of wood wool are that these have excellent insulating property, good durability, and reasonable fire resistant and good impact resistance. The limitation of wood wool is that it needs to protect from moisture as it can absorb moisture and swell and lose its insulating property to some extent.

4.1.4 Sound Proofing Materials

Sound insulation refers to the method of making a part of a building or a room sound proof or to minimize the transfer of sound or noise between the interior and outside of a room. Sound insulation is important to ensure good acoustics in the building by suppressing or eliminating undesired sounds. Acoustically sound buildings add to the comfort of living in the residential complexes and also enhance the job efficiency at work places. However, there is a small difference in sound insulation and acoustics. Sound insulation is controlling/minimizing noise disturbances but acoustics refer to controlling and correcting the sound related flaws. The sound in a building travels through the air or through openings in the members say walls and reflects from slabs or through cracks, if any, in the walls causes echo. Undesired sound or high levels of noise not only cause unpleasant conditions to the occupants but can be even more serious by causing mental stress, fatigue, deafness or even neurological disorders.

4.1.4.1 Characteristics of Good Sound Proofing Material

A good sound proofing material should possess some basic characteristics as discussed below. However, in reality, the heat and sound insulating materials are selected in collaboration and many materials which have good thermal and acoustic control features are preferred at planning, designing and execution levels. The features of good sound insulating material are:

- The texture of sound insulating material should be porous so that it can absorb sound waves and reduce echo.
- The density of sound insulating materials should be low to ensure its less self weight.
- The sound insulating material should be moisture resistant.
- It should be able to withstand termite or insect attack.
- It should be fire resistant.
- The look of the insulating material should be good such that it does not interfere with the appearance of the structure.

4.1.4.2 Common Sound Proofing Material

There are a variety of materials available which can be successfully used for sound insulation of a building either by absorbing sound or by repelling sound to get adequate insulation. Some of commonly used sound insulation materials are:

- **Cellular Concrete:** The cellular concrete is a specially engineered concrete made up of cement, sand, fly ash, water and pre-formed foam. In hardened foam, it has a low density. Its special characteristics include light weight, self compacting, and very good thermal and sound insulation due to air pockets, fire resistant, termite resistant, environment friendly and cost efficient. This material is very popular for insulation purpose.
- **Asbestos:** Asbestos for sound insulation is produced in the form of asbestos papers, cardboards, corrugated sheets and asbestos fibre moulded products. The advantage of asbestos products is that these have good sound as well as thermal insulation and are mostly incombustible. The limitation of asbestos is that it absorbs moisture and then ceases to offer any insulation. Asbestos is also applied in the form of a spray which offers sufficient insulation and protects the metal parts of the building from corrosion.
- **Mineral Wool:** As explained in thermal insulating material, mineral wood is very good sound insulating material also. It can be used for sound proofing of walls and ceilings. Mineral wood absorbs the sound well and is easy and economical to use. It can be filled in cavities such as stud wall or can be used directly on surfaces. Mineral wood is popular for both residential as well as commercial building as it serves both, as thermal and sound insulation material.
- **Acoustic Plaster:** Acoustic plaster to control sound transfer from room to room has been in practice since long. This plaster can be made up of a variety of materials like gypsum, board, paints etc. The plaster layers reduces sound reflections and resonance from the surface and equalizer the sound effects inside the room. These are generally used for public buildings like museum, theatres, dining areas, libraries etc.
- **Porous Tiles:** Porous tiles act as creating a grid system of acoustic tiles fitted directly on walls or on ceiling. Acoustic porous tiles are available in variety of sizes, depths, materials and designs.

These tiles create a finished ceiling with a void above in which the air-conditioned or HVAC ducting and fitting can be located. This method results in a large surface area for absorption of sound and is very effective.

- **Polyurethane Foam (PUR):** Like thermal insulating property, polyurethane foam has good sound insulating feature. The foam is used to reduce sound leakage and is preferred for places like home theatres, gaming rooms, recording rooms etc. The PUR foam improves the overall acoustics of the room. Theses foams are available in a wide range of thickness, sizes and colours to ensure a pleasant look in the room.

Besides, all these materials, a variety of other products like curtains, furniture, carpets, fabric, fibre channels, floor mats etc. are used to create an additional membrane to absorb sound waves inside the room and help in preventing sound transfer from room to room.

4.2 FIBRES

By definition, fibres are a flexible, hair like strand whose length is much more than its width or thickness. Fibres in construction are mainly used as an additive to binding materials. The purpose of adding fibre is to control cracking due to plastic shrinkage and drying shrinkage. When added to concrete, fibres *reduce permeability of concrete* and hence reduce bleeding. Further, some of the fibres in concrete are added to improve resistance to impact and abrasion. The concrete with fibres is called fibres reinforced concrete (FRC).

4.2.1 Types of Fibres

Fibres are broadly classified into two categories:

(a) **Natural Fibres:** These are naturally occurring vegetable fibres. Some examples of natural fibres are jute, bamboo, sisal, flex, coir, banana etc.

(b) **Synthetic Fibres:** Artificially manufactures fibres are referred as synthetic fibres. These include steel fibres, glass fibres, carbon fibres, polymers i.e. plastic fibres.

4.2.2 Jute Fibres

Jute is a naturally occurring fibrous material which is traditionally used to make sacking bags and as a backing material in carpets. Jute fibres are added in concrete mainly to improve its resistance to shrinkage. Experimentation have shown that though the addition of jute fibres causes some adverse effect on the properties of fresh concrete like reducing the workability of concrete but the strength and durability properties of concrete improves with addition of jute fibres. Another advantage of jute fibres is that these are chemically inert and do not alter the hydration of cement in concrete. Jute fibres have a good bonding with concrete. Jute reinforced concrete shows better resistance to abrasion and impact also.

4.2.3 Glass Fibres

The glass fibres are available in the form of strands which can be cut into desired lengths and used as reinforcement in concrete. The glass fibres reinforced concrete (GRC) is lightweight and has very good resistance to aggressive environment like alkali attack, acid attack etc. The ratio and type of glass fibre used depends upon the targeted application. Glass fibres, in general, improve the tensile and compressive strength of concrete. Glass fibres also control the shrinkage cracks in concrete. The use of glass fibres is very popular to make very thin elements with good tensile strength. Glass fibres reinforced concrete panels may reduce the self weight and thickness of concrete by as good as 10 times as compared to steel reinforced concrete panels. The only disadvantage associated with GRC is that it is costlier than convention concrete and requires skilled labour to handling. Glass fibres reinforced concrete is used for renovation works of building, in making of bridge and tunnel lining panels, for water drainage works, as architectural cladding and also as acoustic barriers. In recent times, the use of GRC has also been successfully tried in concrete pavements.

4.2.4 Plastic Fibres

Plastic fibres are typically synthetic fibres which are engineered for concrete to withstand long term alkaline environment. Plastic fibres are manufactured polymer based materials such as polypropylene, nylon or polyethene. Sometimes, even asbestos fibres are used in concrete but the same is not recommended due to the associated health hazards. Plastic fibres are added during the mixing operation of concrete. The advantages of using plastic fibres in concrete include control of plastic shrinkage during fresh or green stage of concrete. The addition of plastic fibres inhibits the growth of micro cracks due to shrinkage by mechanical blocking action of these synthetic fibres. Further, uniformly distributed plastic fibres in concrete discourage the development of capillary channels in concrete and prevent bleeding of concrete. Plastic fibres reinforced concrete has the benefits of increased toughness, improved impact resistance, better tensile strength and increased energy absorption. Another reason which makes addition of plastic fibres in concrete more popular is that the use of these synthetic fibres is compatible with the use of chemical admixtures, pozzolana, mineral admixtures like fly ash, silica fume and metakaolin etc. Basically plastic fibres act as a secondary reinforcement in concrete, improving overall performance and durability of the concrete.

4.3 GEOPOLYMER CEMENT

The term Geopolymer was first used in 1978 by French chemist Joseph Davidovits. The technology behind Geopolymer cement is based on the idea of using inorganic materials with binding properties to produce a product having cementitious characteristics. The production of Geopolymer cement does not require the use of natural resources and hence, does not cause CO_2 emissions into the atmosphere. In Geopolymer concrete, ordinary cement is not used as a binding material, rather, fly ash, silica fume or ground granulated blast furnace slag (GGBS) along with some suitable alkali solutions are used as binders. All these products are industrial wastes whose utilization makes Geopolymer concrete an eco-friendly material. Further, the challenge of safe disposal of these wastes products is also resolved protecting the land and water bodies from pollution.

4.3.1 Constituents of Geopolymer Cement

The major constituents of Geopolymer concrete are coarse aggregates, fine aggregates, a mineral admixture which gets activated in presence of alkaline activators and chemical admixture (if required). The mineral admixture like fly ash, silica fume or GGBS is mixed in powdery form with an alkaline activator to produce a paste which sets and hardens quickly. The commonly used alkaline activators are sodium or potassium hydroxide. These are used in combination with sodium silicate or potassium silicate solutions. Sodium hydroxide (NaOH) and sodium silicate (Na_2SiO_3) are preferred because of the associated economy and higher rate of geo-polymerization. The gel produced by geo-polymerization of these industrial wastes binds the aggregates and the non-reacted material forms the Geopolymer concrete.

4.3.2 Properties of Geopolymer Concrete

Though the properties of Geopolymer concrete depends upon the source of waste material or the type of inorganic material being used to produce it, some of the common properties of Geopolymer concrete are as follows:

- Geopolymer concrete has better workability due to replacement of cement by finer material like fly ash or silica fume or GGBS particles. However, increase in NaOH and Na_2SiO_3 solution reduces the flow of mortar. The use of super plasticizer can increase the workability.
- Geopolymer concrete may have decreased early strength, but it has better ultimate compressive strength.
- Use of geopolymer concrete reduces drying shrinkage as its heat of hydration is less.
- Geopolymer concrete has excellent resistance against aggressive environments.
- The bond strength of GPC is very high, almost four times than that of OPC.
- The tensile strength of GPC is better than that of OPC.

4.3.3 Uses of Geopolymer Concrete

Geopolymer concrete is gaining popularity and has many applications. These include use in:

- **Pavements:** Light pavements can be cast having Geopolymer concrete. The advantage of reduced bleeding in GPC is utilized in making pavements by GPC.
- **Retaining Wall:** Precast panels made up of GPC can be used to raise retaining wall for residential complexes.
- **Water Tanks:** Cost in-situ water tanks for residences can be constructed with Geopolymer concrete.
- **Precast Industry:** Structural members like beams, decks, panels or posts can be made with Geopolymer concrete as they have the advantage of high production in short duration and less breakage during transportation.

UNIT SUMMARY

- Introduction to special construction materials
  - Water proofing compounds
  - Termite types– subterranean termite and non-subterranean termite
  - Thermal and sound proofing materials

- Water proofing materials
  - Admixture
  - Impregnating materials
  - Surface film
  - Sealants
- Termite proofing treatment
  - Pre construction treatment– site preparation, soil preparation, structure barrier
  - Post construction treatment– chemical treatment
- Characteristics of thermal insulating material
  - Fire resistant
  - Resistant to moisture attack
  - Resistant to insect attack
  - Resistant to physical changes
- Thermal insulating materials
  - Mineral foam
  - Polystyrene foam
  - Polyurethane foam
  - Cellulose fibres
  - Wood wool
- Characteristics of sound insulating material
  - Porous
  - Fire resistant
  - Termite and insect resistant
  - Moisture resistant
- Sound proofing materials
  - Asbestos
  - Mineral wool
  - Porous tiles
  - Polyurethane Foam Types of Flooring Tiles
- Fibres used in concrete
  - Types of fibres– Natural and Synthetic fibres
  - Natural fibres– Jute, bamboo, sisal, coir etc.
  - Synthetic fibres– Glass, carbon, plastic fibres
- Geopolymer concrete
  - Constituent of geopolymer concrete
  - Properties of geopolymer concrete
  - Uses of geopolymer concrete

EXERCISES

Multiple Choice Questions

1. Which of the following material is used for damp proofing course?
   (a) Sand (b) Tar (c) Rubber (d) Lead
2. Water proofing layer __________ the process of water absorption.
   (a) Slows down (b) Increases (c) Does not effect (d) Cannot say
3. Water proofing compounds are available in the form of
   (a) Powder only (b) Powder and paste form
   (c) Powder, paste and liquid form (d) Powder, paste, liquid and gaseous form

4. The other name for termite is
   (a) Beatles (b) Bugs (c) Insects (d) White Ants
5. Which of the following chemical is used for soil treatment in termite proofing?
   (a) Chlorine (b) Bromine (c) Aldrin (d) Potassium hydroxide
6. Which termites cannot survive without maintaining a contact with the soil?
   (a) Dry wood (b) Subterranean
   (c) Non Subterranean (d) All types of termite
7. Which of the following is not effective for termite treatment?
   (a) Emulsion (b) Chemical barrier (c) Metal barriers (d) Soil treatment
8. The transfer of heat within a material is known as
   (a) Conduction (b) Radiation (c) Convection (d) Propagation
9. A good thermal insulating material should have __________ thermal resistance.
   (a) *Low* (b) *High* (c) *Moderate* (d) Cannot say
10. The presence of moisture in thermal insulating material __________ its thermal insulation performance.
    (a) Improves (b) Decreases (c) Does not effect (d) Cannot say
11. Which of the following is NOT a characteristic of fibre reinforced concrete?
    (a) High strength (b) Light weight
    (c) Sensitive to temperature change (d) Durable
12. Which material is used for making of lining of a sound proof chamber?
    (a) Polystyrene (b) Wrought iron (c) Titanium (d) Stainless steel
13. Mechanical properties of fibre-reinforced composites depend on
    (a) Properties of constituents (b) Fibre length
    (c) Fibre orientation and volume fraction (d) All of the above
14. Which of the following is an example of natural fibre?
    (a) Asbestos (b) Plastic (c) Glass (d) Jute
15. Which metal is used to create metal barrier along the walls to prevent termite attack?
    (a) Copper (b) Iron (c) Titanium (d) Anyone
16. The biggest challenge associated with cement production is
    (a) Energy consumption (b) CO_2 emission
    (c) Raw materials (d) Transportation cost
17. Which of the following combination represents an alkaline activator solution for geopolymer concrete?
    (a) Alkali silicate and carbonates (b) Alkali hydroxide and carbonates
    (c) Alkali hydroxide and silicates (d) Alkali hydroxide and water

18. The early strength of GPC is __________ than that of ordinary concrete.
(a) More (b) Less (c) Same (d) Cannot say

19. Cavity walls is a solution for
(a) Water proofing (b) Thermal insulation
(c) Sound insulation (d) All of the above

20. Asbestos fibres are not popular in usage because of associated.
(a) Health hazards (b) Cost
(c) Difficulty in handling (d) Non availability

ANSWERS																			
1.	(b)	2.	(a)	3.	(c)	4.	(d)	5.	(c)	6.	(b)	7.	(c)	8.	(a)	9.	(b)	10.	(b)
11.	(c)	12.	(a)	13.	(d)	14.	(d)	15.	(a)	16.	(b)	17.	(c)	18.	(b)	19.	(d)	20.	(a)

Short Answer Type Questions

1. Why is water-proofing necessary for a building?
2. Differentiate between acoustics and sound insulation.
3. How do you classify termite?
4. State the advantages of geopolymer concrete over ordinary concrete.
5. What are the benefits of adding fibres to concrete?
6. Discuss the applications of geopolymer concrete.
7. Enlist various materials used for sound insulation.
8. State the characteristics of good thermal insulating material.
9. What is mineral wool? Explain its usage for insulation purpose.
10. How do sealants help in water proofing?

Long Answer Type Questions

11. Discuss various techniques adopted for water proofing of structures.
12. Explain the post construction treatment for termite in detail.
13. What are various materials used for thermal insulation in buildings? Explain in detail.
14. What is geopolymer concrete? Discuss the manufacturing of geopolymer concrete. Also state its properties.
15. Write an explanatory note on use of plastic fibres in concrete.

PRACTICAL

Experiment-1

Aim: Apply the relevant termite chemical on given damaged sample of timber.

Materials Required: Sample pieces of damaged wood, cloth for dusting, brush/spraying bottle/injection, thinner, turpentine oil or anti termite chemical.

Theory: There are various methods to treat damaged wood against termite attack. These methods include use of anti termite chemicals on damaged piece of wood with the help of paint brush or by spraying the chemical on the surface of wood or injecting the chemical into the wood for better penetration or even fumigation i.e., exposing all termite inside the wood to fumes of some strong anti termite chemical like sulphuric fluoride. The aim of all these applications is basically to create a barrier to prevent survival and spread of the termite.

Procedure:

1. Take the given damaged piece of wood.
2. Observe the sample carefully to notice the damaged parts.
3. Remove all the dust from the surface of sample wood with the help of a cloth.
4. Wipe the surface clean before applying any anti termite chemical for treatment. Thinner or turpentine oil is generally used to clean the surface.
5. Select a suitable method to apply a layer of anti termite chemical. It can be achieved by brush or spray or by injection.
6. Create a layer of anti termite chemical on the surface of damaged wood by brush or spraying. If the damage is deep inside the wood, inject a coat of chemical into the wood.
7. Allow it to dry for at least 12 hours.
8. Apply more coats, if necessary.

Observations:

1. Record the pictures of damaged part of wood.
2. Record the pictures of wood after application of anti termite treatment

Result: Anti termite treatment can protect the damaged wood from further deterioration.

Precautions:

1. Proper cleaning of wood surface is necessary before applying any treatment.
2. The surface should be properly dried before Appling next coat on the surface for anti termite treatment.

KNOW MORE

Water is one of the most destructive harmful weathering elements for a structure. It is always costlier to repair a water effected structure than to make it water proof since its construction. Also, it is noteworthy that roof or terrace is the most common element of structure to cause some water leakage related damage. Hence, efforts must be taken to make it water seepage proof at the design stage itself. Water proofing techniques involve some additional cost at the time of construction but it is worth spending it as the water proofing techniques adopted later are neither cost effective nor efficient. Sometimes, inadequate workmanship, poor quality materials or water logging may be the reasons for dampness in a structure.

Termite attack on buildings is the other most common reason for damage to a structure. Termites have the potential to cause damage of a significant level to the structure. Termite proofing is, therefore, essential to ensure longevity and durability of a structure. Termite proofing can be ensured more efficiently if done before or during construction rather than anti termite treatment on effected structure.

The other cause of discomfort for inhabitants of a building is transfer of heat and sound, in and out, from the structure. Thermal and sound insulation becomes more important for commercial places like factories, manufacturing units etc. where excessive temperature changes and long time exposure to loud sound may cause health hazards to the workers. The selection of thermal and sound insulation materials should be done to ensure that those materials are safe, durable and effective for a broad range of applications. It will not be wrong to state that thermally well insulated structure with good acoustics is energy efficient and ecologically balanced structure, offers a pleasant stay to the occupants.

Another newly developed material, geopolymer concrete is considered as an innovative and eco-friendly construction material. Ordinary Portland cement (OPC) being on energy intensive material, has always been under surveillance. Further, ordinary cement production leads to very high carbon-dioxide (CO_2) emission. A lot of research has been conducted to find a substitute to cement and geopolymer concrete is a result of all these researches. Geopolymer concrete has the advantages of high compressive strength, high tensile strength, low creep, low drying shrinkage, fire proofing, highly durable and better resistance to chemical and environmental attacks. The limitations associated with GPC include requirement for special handling techniques to use chemicals like $NaOH$ and Na_2SiO_3 which are harmful to human beings to some extent and needs care while handling. The high cost of alkaline solution is another barrier in the use of geopolymer concrete.

REFERENCES AND SUGGESTED READINGS

- Duggal, S. K, Building Materials, New International, New Delhi.
- S.K. Sharma, Civil Engineering Construction Materials, Khanna Publishing House, Delhi

5

UNIT SPECIFICS

This unit specifies the following aspects:

- Knowing the special processed construction materials like geo-synthetics, artificial sand, artificial timber etc.
- Understanding the constructions of Plaster of Paris and its uses.
- Knowing various types of points and their specific uses.
- Understanding the use of industrial waste materials in construction industry.
- Knowledge on agro waste materials.

Besides the above stated aspects, the practical consideration on the applications of specially processed construction materials has been covered in this unit. The use of paints and plaster of paris is inevitable from aesthetic considerations for a structure. The chapter elaborately covers the constituents and uses of paints, varnishes that are going to build sufficient knowhow amongst users and help them choose an appropriate paint for a specific application. Further, emphasis has been laid to understand the use of industry waste products and agriculture by-products in construction.

At the end of the unit, a number of objective multiple choice questions have been given to assess the understanding of topics. Besides, short and long answer type questions have been designed on the basis of bloom's taxonomy to prepare the students for examination. The knowmore section at the end of the unit gives some additional insight on the usage of special processed construction materials. Some QR codes have been generated within the chapter to give a deeper knowledge on processed construction materials.

The practical's related to application of paints and preparations of mortar using artificial materials have been explained at the end of the unit. The practical's will build a confidence on real time applications in the students as the students will perform the exercises actually in the laboratory. The whole chapter has been prepared with almost care to develop an iniquities approach amongst students to understand the usage of processed construction materials.

RATIONALE

The unit processed construction materials that help the users to understand the use of non-conventional and specially processed materials, manufactured from industrial or agricultural waste, in the construction industry. The constituents and applications of processed materials in construction have been explained thoroughly in this chapter. Some of the industrial by-products, whose disposal itself is a challenge, have been explained here for their potential usage for the construction purposes. Besides, discussion on specially designed materials like geo-synthetics and ferrocrete generate a comprehensive knowledge bank on construction material for the students. The advantages and limitations on the use of various types of paints, varnishes and Plaster of Paris have been explained to ensure appropriate selection of these products from a wide range of products available.

Also, the knowhow on some of the processed materials for construction provides the readers to look for or generate more alternative materials for construction based on industrial and agricultural by-products. A rationale on the use of processed materials in construction may be sought by users after understanding this chapter thoroughly.

PRE-REQUISITE

Knowledge on all types of natural, artificial and special construction materials (i.e., Unit-II, Unit-III and Unit-IV)

UNIT OUTCOMES

Following is the list of outcomes from unit on Processed Construction Materials:

U5-O1: Define the application of Plaster of Paris, paints and varnishes.

U5-O2: Understand the need for industrial and agro based waste materials in construction.

U5-O3: Know various industrial waste materials and their usage in construction.

U5-O4: Know various agro waste materials and their usage in construction.

U5-O5: Know various special processed construction materials.

Unit-5 Outcomes	EXPECTED MAPPING WITH COURSE OUTCOMES (1- Weak Correlation; 2- Medium Correlation; 3- Strong Correlation)				
	CO-1	CO-2	CO-3	CO-4	CO-5
U5-O1	1	1	1	1	3
U5-O2	1	2	2	2	3
U5-O3	1	2	2	2	3
U5-O4	1	2	2	2	3
U5-O5	1	2	2	2	3

5.1 PLASTER OF PARIS (POP)

Plaster of Paris (POP) is a white powdery substance available in common market and used for getting finished smooth surfaces in construction. Also, it is commonly used as sculpting material and by orthopedist for gauzing bandages for treatment of fracture of bones. In chemical terms, POP is hydrated calcium sulphate which is obtained from calcining gypsum. Plaster of Paris is obtained by heating gypsum at a high temperature (say 373 – 393K). Plaster of Paris is also known as gypsum plaster. The chemical formula for POP is $CaSO_4.2H_2O$.

5.1.1 Properties of Plaster of Paris

The important properties of plaster of paris are as follows:

(i) POP is usually available in powdery form and is white in colour.
(ii) When water is added to gypsum, solid crystals are formed.
(iii) The setting of plaster of paris in an exothermic reaction. The reaction can be catalyzed by sodium chloride whereas it can be retarded by addition of alum or borax powder.
(iv) At high temperature, say of the order of 473 K, dead burnt plaster of paris is formed.
(v) Plaster of paris can be shaped into desired size and shape as required.

5.1.2 Types of Plaster of Paris

Some of the types of plaster of paris are as follows:

(i) **Gypsum Plaster:** This is the most common form of POP being used in modern days. Gypsum plaster is basically white powder form of calcium sulphate hemihydrates, formed by heating gypsum at a temperature of 120°C to 180°C.
(ii) **Clay Plaster:** This type of plaster was used in ancient times in early nineteenth century. Clay plaster is a mixture of clay, sand and water which is added with plant fibres to improve its tensile strength. This type of plaster has been popular in interior decorative works in houses.
(iii) **Lime Plaster:** Lime plaster is a mixture of calcium hydroxide and sand. When this plaster gets exposed to carbon dioxide of atmosphere, it starts to set and calcium hydroxide gets transformed to calcium carbonate. To prepare lime plaster, limestone (calcium carbonate) is heated at a temperature of 850°C and quick lime (calcium oxide) is produced. When water is added to it, slaked lime (calcium hydroxide) is produced which can be stored either in a powder form or as a wet putty. When more water is added, it forms a paste, called lime plaster, which is used as a building material and also for making mural paintings.
(iv) **Cement Plaster:** A mixture of plaster, cement, sand and water is called cement plaster. It is mainly used in masonry to obtain a smooth surface. Cement plaster has a good strength, hardness, quick setting and durability.

5.1.3 Uses of Plaster of Paris

The uses of plaster of paris mainly include:

(i) Use in dentistry to make casts.
(ii) Use in hospitals to set fractured bones or to treat sprains.
(iii) Used to make toys, decorative materials like statues, show pieces, or even ornaments.
(iv) Used as a fire proofing material.
(v) Used in sealing air gaps in apparatus used in chemistry labs.
(vi) Used to make blackboard chalks.

5.2 PAINTS

Paint is a substance which is applied over a surface of metal or structure or wood and is left for drying to give a thin, decorative and protective coating or covering. Paints are available in a variety of colours.

Paints provide a smooth finish to the surface on which these are applied. The selection of particular types of paint depends upon the nature of surface and the extent of finish required for a particular job.

5.2.1 Constituents of Paint

Paints mainly consist of following four elements:

(i) **Pigment:** The main function of pigment is to provide colour and gloss.

(ii) **Resin or Base:** The function of resin is to act as a binder to hold the pigment particles together and to provide its adhesive to the surface.

(iii) **Solvent or Vehicle:** The function of solvent is to act as a carrier for the pigment and resin and to allow the easy spread of paint on the surface. The solvent may be organic or inorganic in nature. Liquids, chemicals and water are some of the examples of solvent. Sometimes, a thinner is added to a paint to get desired consistency of paint to make it easy to apply. Turpentine oil, white spirit, alcohol are some of the common thinners.

(iv) **Additives:** Additives are special compounds which are added to paint to enhance some of its properties like ease of brushing or spraying, resistance to scuffing, sagging and ease of drying etc. A substance which acts as a catalyst and accelerates the process of drying of paints is called drier.

5.2.2 Function of a Paint

The common purposes for which paints are applied to concrete/cement surface or metals are as follows:

(i) **Decoration:** Paints are applied on surface to improve the beauty of the material. Paints are available in wide variety of colour shades and gloss. Paints enhance the aesthetic appeal of a surface.

(ii) **Protection:** Paints act as a protective layer on surface against the action of weathering, attack by insects or oxidation process. Paints on metallic surface prevent corrosion of the surface.

(iii) **Identification:** Paints can be used to mark various elements of a building differently. Paints can be used to make varied designs, patterns on the surface. Hence, paints may use as an identification material.

(iv) **Sanitation:** Paints facilitate the cleaning process of the surfaces. Most of the paints are easy to clean either by washing or by wiping with cloth.

5.2.3 Characteristics of a Good Paint

A good paint should have following features:

(i) A good paint should be able to cover maximum area with minimum paint. It should be easily spreadable. The spreading capability of paint is also dependent on the absorbing power of the surface. Cement and plaster are more absorbent as compared to metals and wood.

(ii) A good paint should be able to adhere well to the surface applied.

(iii) The paint should dry quickly (say within 24 hours) and should leave an impervious, uniform, smooth, hard and wear resistant layer on the surface.

(iv) A good paint should not fade due to environmental conditions or with the passage of time.

(v) A good paint should be economical and affordable.

(vi) A good paint should be easy to clean and attractive in looks.

(vii) A good paint should be inert i.e., should not cause any harm to the surface.

(viii) A good paint should have good resistance to weathering actions of heat, water and wind etc.

5.2.4 Types of Paints

Paints are broadly classified into two categories:

(i) Water based paints. (ii) Oil based paints

(i) **Water Based Paints:** Water based paints are the most common types of paints. Such paints consist of pigment, binder, drier and mainly water as solvent. Only water is added to these paints to make it ready to use. The advantage of water based paints is that is requires lesser time than oil paints to dry. Water based paints are easily washable on drying and maintain their original colour for a longer period. Water based paints are generally cheaper and are suitable for interior as well as exterior walls. Water based paints are further classified into two categories as:

(a) **Distemper:** Distempers are one of the ancient types of paints. Distemper consists of chalk, pigment and water bound with either some animal glue or an adhesive. Distemper has the advantage of being inexpensive and providing a cost-effective painting solution for interior walls. Most of distempers are light in colour and provide good reflective coating. Distemper has the limitations of not being durable and not suitable for exterior walls. Distempers are available in two types namely, acrylic distemper, which is the highest quality distemper providing a pleasant smooth matte finish and synthetic distemper, which is water based low finish economical distemper.

(b) **Emulsion:** Emulsions are much superior to distemper. These are formed by mixing chalk and pigment with water and an emulsifying agent to prevent separation of combination. Emulsions have higher coverage capacity than distemper. They also have the advantage of ease of application, easily washable and an overall pleasant look/finish. Emulsion can be used for interior as well as exterior finish. Modified acrylic type of emulsions is generally preferred for exterior applications.

(ii) **Oil Based Paints:** Oil based paints are made by adding either natural (Linseed) oils or synthetic (alkyd) oils as solvent. Oil based paints are commonly known as enamel paints. These are available in a wide variety of colour, shades (dark and light) and finish (glass, matt and satin). Oil based paints are much more durable than water based paints. However, the oil based paints take longer to dry and require turpentine oil or thinner for cleansing. The colour of oil based paints also fades away with time. When these are to be applied to iron, a coat of red oxide is applied first to protect iron from getting rotten. Oil based paints are used for all types of works like walls, ceiling, doors, windows, iron frames, iron meshes, furniture, wooden frames and even art and craft items etc.

5.2.5 Cement Paint

Cement paint consists of boiled linseed oil to which dry cement is added in good quantity (say 65 – 75%) and stirred to get the consistency of paint. The cement paint is available in powdery form in a limited range of colours. It is a water based paint which has to be used within two hours of its preparation. It dries up easily and hardens up like a rock. Cement paint has following important properties:

(i) It possesses better water proofing qualities and hence can be used for exterior as well as damp places.

(ii) It has good strength, hardness, density and good durability. It prevents even the growth of fungus and bacteria on exterior surface.

(iii) It requires ample amount of water and provides good coverage area and hence is economical.

Cement paint can also reduce the collection of dust on surfaces and is easily washable. Cement paint has gained popularity as the last finish coat for exterior surfaces. It is extensively used for garages, driveways, walls and ceiling of residences, offices, workshops, public building, hotels, warehouses, factories and even for concrete furniture in parks.

5.2.6 White Wash

White wash is the process of providing a clean, neat and uniform finish to the surfaces of buildings. White wash is prepared as a mixture of pure fat slaked lime into sufficient quantity of water (5 litres of water for every 1 kg of lime) and then screened through coarse cloth. A mixture of boiled gum or rice water is added in white wash in small proportions to act as binder. This solution is called white wash. The surface is thoroughly cleaned off all dust, dirt or any other foreign material before the white wash is applied on it. White wash is applied by brushes for a specific number of coats, usually three on each surface. Sometimes a colouring pigment, which does not react with lime, is added in white wash to get desired shade on the surface. White wash is much inferior but economical than paints.

5.2.7 Varnish

Varnish is a solution of a resinous substance in oil or alcohol. Varnish is more or less transparent and provides a transparent, hard, protective and glossy film on the surface. Varnishes are generally provided over wooden unpainted surfaces to maintain its original look of natural grains and fibres. Varnish is often added with a drier to accelerate the drying process. Resin, solvent and drier are the main ingredients of a varnish.

Certain precautions should be taken care of while varnishing the surfaces. These include that varnishing should not be carried out on a rainy day and it should become surface dry in 6 – 8 hours. Special kinds of fine haired brushes are used for varnishing. Varnishing is generally recommended for interior works and painting is preferred for exterior works. Another associated term with varnish is 'Polish'. Polishes are thin varnishes with spirit to allow a smooth, fast working and finish.

5.2.7.1 Objectives of Varnishing

The surface is applied with varnish with the following objectives:

(i) **Ornamental Look:** Varnishing enhances the overall look of the surface. It brightens up the appearance of grains and fibres in unpainted wooden surface.

(ii) **Protection:** Applying varnish over wooden surfaces of doors, windows, furniture etc. provides a protective film against atmospheric actions and increases their lifespan.

(iii) **Glossy Appearance:** When varnish is applied on painted surface, it provides a glossy attractive film on the surface.

(iv) **Durable:** Varnishing on painted surfaces provides a protective covering and increase the durability of the paint coat.

5.3 INDUSTRIAL WASTE MATERIALS

There are many industrial waste materials which have been used for construction purposes successfully. These by-products from industries are otherwise waste materials and have generally challenges associated with their safe disposal. The use of industrial by-products in construction industry has been tired through worldwide researchers and some of these wastes have given fruitful results. The utilization of industrial waste products in construction addresses the problem of their safe disposal and also enhances the quality of infrastructure. Some of the popular industrial wastes used in construction are flyash, blast furnace slag, marble and granite waste and recycled aggregates etc.

5.3.1 Flyash

Flyash is a by-product obtained from modern thermal power plants where powdered coal is burnt in presence of air. Flyash consists of fine particles of burnt fuel. Flyash resembles cement in colour and fineness. Flyash contains good percentage of silica, calcium oxide and alumina and can be safely added with cement or clay to enhance their certain properties. Flyash when added in concrete reacts with free lime and produces binder which results in strength improvement in concrete. Flyash can be used in construction industry for many purposes like manufacturing of bricks, manufacturing of cement as an additive or substitute to cement or many more. Broadly flyash is classified into two categories:

- **Type C:** Type C flyash is produced by burning of sub-bituminous coal containing lime more than 10%. It has cementitious as well as pozzolanic properties.
- **Type F:** Type F flyash is obtained by combination of bituminous coal containing lime less than 10%. Type F flyash has pozzolanic properties.

5.3.1.1 Advantages of Using Flyash

(i) The use of flyash in concrete reduces the rate of hydration of cement. Though, this result in reduction in early strength of concrete but the ultimate strength of flyash concrete is better than ordinary concrete. Reduced rate of hydration results in low heat of hydration and hence less thermal cracking.

(ii) The addition of flyash in concrete improves workability. The particles of flyash are spherical in shape and finer than cement particles. This reduces the friction between aggregates and creates a lubricating effect, increasing workability.

(iii) The reaction between flyash and lime gives additional C-S-H gel in concrete, resulting in improved resistance to sulphate attack.

(iv) The use of flyash reduces the available free lime and prevents the corrosion of reinforcing steel bars.

(v) The flyash concrete has less permeability and hence improved water-tightness which provides better resistance against carbonation.

5.3.1.2 Uses of Flyash in Construction

(i) **Portland Pozzolana Cement:** Flyash is blended with Ordinary Portland Cement (OPC) to produce Portland Pozzolana Cement (PPC). The proportion of flyash in PPC can vary between 10 to 30%.

(ii) **Flyash Concrete:** Flyash in concrete is used as a replacement to Portland cement or as an additive material. The recommended percentage of flyash in concrete is upto 30%, based on the research.

(iii) **Flyash Bricks:** Flyash is added with clay and then heated in kiln at about 1000°C to get flyash bricks. The use of flyash in brick making is increasing these days. Flyash upto an equal amount of clay may be suitably used in brick manufacturing.

(iv) **Soil Stabilization:** Soil stabilization refers to permanent improvement in physical and chemical properties of soil by addition of mineral or chemical additives. Flyash is added in soil as a mineral additive for its stabilization.

(v) **Embankments:** Flyash is used for construction of embankment. Besides, flyash is also used as a flow able fill and as a component in geopolymer.

5.3.2 Blast Furnace Slag

Slag is a non-metallic by-product produced from manufacturing of pig iron and steel. Blast furnace slag in granular form is produced by quick chilling of slag under water. However, slag can be converted into spongy or pumice structure by foaming i.e., by running slag under a limited amount of water. The heat of the furnace slag converts the water into steam which makes the slag swell and results in a spongy look. Blast furnace slag predominantly consists of CaO, SiO_2, MgO and Al_2O_3 and hence can be used with cement or concrete.

5.3.2.1 Advantages of Blast Furnace Slag

Slag has the advantage of its latent heat which can be utilized to save energy, save natural resources and for environment conservation in construction related activities. The advantages of blast furnace slag can be attributed to the greater mobility characteristics of its particles. The advantages of blast furnace slag include:

(i) The colour of granulated blast furnace slag is very even and uniform and it gives a better surface finish and pleasant look.

(ii) The heat of hydration of blast furnace slag added concrete is less which results in less thermal cracking and less early strength but the ultimate strength is same or higher.

(iii) The slag added concrete has better resistance to alkali and chemical attacks.

(iv) Slag reduces the chances of efflorescence.

(v) The slag added cement and concrete are chemically more stable and hence more durable.

5.3.2.2 Use of Blast Furnace Slag

Blast furnace slag has following applications in the field of construction:

(i) Portland Blast Furnace Slag Cement: This cement is produced by grinding ordinary Portland cement clinker and dry granulated slag with a small proportion of gypsum. The proportion of slag can be kept between 30 to 50% of cement. Blast furnace slag cement has lower early strength due to reduced rate of hydration but the ultimate strength is higher or equal to that of OPC. This cement possesses properties related to low heat of evolution and has high resistance against water. Blast furnace slag cement can be used in water retaining structures.

(ii) Light Weight Concrete: The blast furnace slag in spongy form has low density. These slag granules can be used as replacement to aggregates in making of light weight concrete.

(iii) Blast Furnace slag in ungrounded form can be used as a base layer material for road construction. It can also be used as a fill in embankments.

(iv) Air cooled slag is used to get dense concrete which is more durable and preferred for concrete pavements construction.

(v) Slag wool or in fibrous form is used for insulation processes.

(vi) Other uses of blast furnace slag include in making of glass, concrete blocks, filtration medium, back filling and for soil stabilization.

5.3.3 Granite and Marble Polishing Waste

Granite and marble are naturally occurring rocks which are popular for their usage in flooring due to high strength, pleasant looks and availability in wide range of shades and sizes. During the processing of granite and marble rocks, huge amount of their powder is generated as a waste material. Since granite powder has not been used much for industrial purposes till date, there is a big deposit available for granite and marble powder. The deposition of this waste powder is a significant environment problem which is causing soil contamination, plant pollination and even hazardous to human health. Many attempts have been made by researchers to utilize granite and marble waste powder for construction purposes. Currently, this waste powder is being successfully used as cement composite and in epoxy flooring.

5.3.3.1 Advantages of Granite and Marble Waste

(i) Granite and marble polishing waste is available in varied sizes which are similar to coarse aggregate size, fine aggregate size and even in powdery form. Hence, this waste can be used as a replacement to coarse as well as fine aggregates in concrete.

(ii) The granite waste has good CBR value which makes it suitable for road pavement construction.

(iii) Huge deposit of granite waste is available. Hence, their usage makes the construction activity economical.

(iv) The use of granite and marble waste in construction is beneficial from economic and environmental point of view and waste results in saving of natural resources.

5.3.3.2 Uses of Granite and Marble Waste

(i) Due to similarity of granite and marble waste with conventional coarse aggregate, these can be used as a substitute in concrete. The proportion of granite waste can be kept in the range of 10% to 40% and it gives better compressive strength than traditional mix.

(ii) Granite and marble coarser waste is used as coarse particulate aggregates in concrete road pavements.

(iii) These wastes are used as aggregates for soil compaction under foundation. The problematic layers of soil under foundation are replaced with granite and marble waste to get proper compaction.

(iv) Granite and marble waste in fine form is used as a replacement to natural sand in concrete. Concrete with granite and marble waste gives better compressive and flexural strength.

(v) Granite waste debris can be used in manufacturing of ceramic tiles. It acts as a secondary flux material and gives long term durability to tiles.

(vi) Also, granite and marble waste is used in making of building precast blocks of concrete.

5.4 AGRO WASTE MATERIALS

Agro waste materials are those materials which are produced as a by-product during agricultural practices. There is a huge quantity of agro waste produced every season whose disposal is a challenge. The agro wastes are organic in nature and hence cannot be used without any treatment for construction purposes. The most common practice for disposal of agro waste is to dry them and then burn it. Burning of agro residue poses a serious environmental threat by causing air pollution which is hazardous to human and plant health. Though a lot of studies have been carried out to analyse the effect of utilization of agro waste in construction industry, but still very rare information is available to build the confidence for the use of agro waste materials in construction. Some of the agro waste used in construction is rice husk, bagasse, coir fibres, corn cobs etc.

5.4.1 Rice Husk

Use of rice husk has been successfully tried in making of cement bricks, blocks and even in concrete. Recycling of rice straw husk for building processes can provide a comfortable solution to the major problem caused by burning of paddy fields. The rice husk has high silica content and is resistant to moisture and fungal decomposition. The addition of chopped dried rice husk improves the mechanic properties of cement brick and strength properties of concrete mix. The fire resistance is also better for rice husk added mixes. Also, rice husk acts as a fibre and improves the shrinkage resistance of the mix. Rice husk added clay or concrete has better thermal and acoustic insulation. Besides, the ash obtained by burning rice husk has pozollanic properties and it is used as a partial substitute to cement in concrete. Rice husk ash is also used in manufacturing of refractory bricks and for improving residual soil properties.

5.4.2 Coir Fibres

In regions close to tropics, coconut occupies a major space in agriculture. Coconut has a lot of fibres around its body in both ripe as well as unripe state. This fibre is used in making of sacks, brushes, doormats, ropes and mattresses. Besides, these fibres have significant uses in the field of building materials. These fibres in dry state are referred as coir fibres. Coir fibres are added to concrete mixes to reduce shrinkage cracks. These are also added to cement soil bricks as reinforcement and improve the tensile strength of the mix. The thermal conductivity for coir fibre based concrete is less and hence it is used as bio insulation in concrete ponds. Coir fibres reinforced cement mortar is laid as roof sheet on structures to reduce heat transfer and save energy. The properties of coir fibre can further be enhanced by treating it with 2% of alkali. Coir fibres have good mechanical strength properties, water proofing properties and more durability as these are resistance to insect or rot attack.

5.4.3 Bagasse

Sugar cane bagasse is an agro based waste generated from sugar and ethanol industry. The by– product is generated in huge quantities and is burnt to produce Sugar Cane Bagasse Ash (SCBA). This ash has been used a fertilizer or is disposed in landfills since ages. The modern researchers have proved that SCBA has pozzolanic characteristics and can be used as a replacement to cement upto some suitable substitution levels. The physical and chemical properties of the bagasse ash is dependent on the temperature on which bagasse is burnt. Melting or burning bagasse at higher temperature (700°C – 800°C) results in spherical particles which improve the fresh properties of concrete. Bagasse ash added concrete has better mechanical and strength properties. Sugarcane bagasse ash has been successfully utilized in production of conventional concrete, high performance concrete and also self-compacting concrete. SCBA can also be used in manufacturing of unburnt bricks and paver blocks. These bricks have almost NO efflorescence. An optimum dosage for cement replacement by bagasse ash is 10% – 20%. Bagasse as such is used as a fibre in concrete but its scope is limited to shrinkage control in concrete but bagasse ash has wide application in construction activities.

5.5 SPECIAL PROCESSED CONSTRUCTION MATERIALS

Processed construction materials are those innovative materials which are manufactured for advanced facilities or specific purposes in building. These are crafted with natural or artificial construction materials to achieve the desired properties or objectives. Some of the commonly used special processed construction materials include geo-synthetics, ferro-crete, artificial timber, artificial sand etc.

5.5.1 Geo-Synthetics

Geo-synthetic is made of geo i.e., earth or soil and synthetic i.e., manmade. Geo-synthetics are manmade materials used to improve the properties of soil. Geo-synthetics are synthetically manufactured petro-chemical based polymers which are biologically inert and do not get decomposed by insect, bacterial or fungal attack. Some of the geo-synthetics used in construction activities are polyester, high density polyethene, polypropylene, ceramics, synthetic fibres and composites.

5.5.1.1 Types of Geo-Synthetics

Depending upon the basic component and the manufactured and final product, geo-synthetics are divided into following types:

(i) **Geo-textiles:** Geo-textiles form the largest group of geo-synthetics. These are textiles consisting of synthetic fibres. Geo-textiles are porous and allow water drainage through it. Geo-textiles can be woven or non-woven or of knit category.

(ii) **Geo-grids:** Geo-grids are plastics having open grid like appearance. Geo-grids are primarily, used as the reinforcement to the soil.

(iii) **Geo-nets:** Geo-nets are materials with open grid like structure formed by polymer strands intersecting at a constant acute angle. These are used to carry large fluid through them.

(iv) **Geo-membranes:** Geo-membranes are continuous flexible sheets made up of one or more synthetic materials. These sheets are mostly impermeable. Geo-membranes are used as liners or barriers for fluids.

(v) **Geo-composites:** When two or more geo-synthetic types combine to form a new geo-synthetic material, it is called geo-composite. For example: Geo-textile – geo-net, Geo-textile – geo-grid etc. These are mainly used for controlling drainage.

(vi) **Geo-others:** Many other types of geo-synthetics such as geo-cells, geo-mats, geo-strips, geo-foam etc. are referred as geo-others. Geo-cells are thick, three dimensional networks made up of strips of polymeric sheets. Multiple layers of geo-cells are combined to form thicker foam like structure called geo-mattress. Geo-mats are 3D permeable mats used for soil protection and erosion control.

5.5.1.2 Functions of Geo-synthetics

The following are the five major functions of geo-synthetics:

- **Drainage:** Geo-synthetic material is used as a drain to carry fluid i.e. liquid and gases through less permeable soils. This feature of geo-synthetics is helpful in forming pavement edge drains, slope drains and abutment and retaining wall drains. Proper drainage through geo-synthetic drains helps in regulating the water logging related issues.
- **Filtration:** Geo-synthetic materials act as filter like sand filter and allow the water to move through the soil. Geo-textiles are used to control the movement of soils into drainage or pipes. Similarly, permeable geo-mats are used to prevent soil erosion.
- **Reinforcement:** Geo-synthetics are used as reinforcement in soil mass to improve its strength and deformation properties. Geo-textiles and geo-grids are commonly used to enhance the tensile strength of soils; Embankments can be constructed over soft soils by providing geo-synthetic material as reinforcement.
- **Separation:** Geo-synthetics act as a separator between two layers of soils, having different particle size distribution and hence different properties. Geo-textiles are used to regulate the mixing of road base materials with underlying soft sub grade soils.
- **Erosion Control:** Geo-synthetic materials are used to reduce soil erosion caused by rainfall or surface water runoff. Geo-synthetic mattress and geo-mats/blankets are placed over the exposed soil on slopes which prevent the soil to be get eroded with water.

5.5.2 Ferrocrete

Ferrocrete is a special processed construction material made by sandwiching one or more layers of steel wire mesh or chicken mesh in cement mortar. Ferrocrete is different than RCC as it has continues well distributed grid reinforcement rather than steel bars. The wire mesh in Ferrocrete results in higher tensile strength, and good ductility resulting in better crack resistance and better impact resistance than conventional concrete. Ferrocrete can be successfully used in construction of buildings, ducts, water reservoirs, foot bridges, manhole covers, drains etc. The other name for Ferrocrete is ferro-cement.

5.5.2.1 Properties of Ferrocrete

The important properties of Ferrocrete are:

- Ferrocrete has low water cement ratio. It helps in producing impermeable structures.

- Ferrocrete has high density which makes it durable against climate attacks for very long period of age.
- Ferrocrete shows very good bonding between cement mortar and iron mesh, making it exceptionally good for resistance against wear and tear.

5.5.2.2 Advantages of Ferrocrete

The major advantages of Ferrocrete are:

- Use of Ferrocrete in construction, makes the structure thin, light and aesthetically pleasant.
- Ferrocrete can be easily used in precast construction.
- Repairs required due to local damage are easier in construction with Ferrocrete.
- Ferro-cement can be used to produce high early strength concrete or mortars.
- Construction using Ferrocrete is economical and sturdy than RCC construction.

5.5.2.3 Disadvantages of Ferrocrete

Some of the limitations on use of Ferrocrete are as follows:

- The construction with Ferrocrete is labour intensive.
- The compressive strength of Ferrocrete is less than that of RCC construction.
- Skilled labour is required to handle rafters and purlins user in logical construction by Ferrocrete.

5.5.3 Artificial Timber

Artificial timber is a relatively new term used for a material, which can be used as a substitute to timber and is made from solid materials like flyash, silica, bituminous sand and biodegradable cellulose ash. It is a true sustainable innovative construction material by Roy Research and Technology. The artificial wood simulates and feels like wood but is manufactured by utilization of waste products like flyash, bottom ash etc. whose disposal is otherwise a problem. Just like wood, artificial timber is manufactured with different colours, textures and appearance of grains. Artificial timber can be manufactured in the form of blocks, beams or panels and can be cut in varied shapes and sizes for use in making of furniture, flooring or even construction.

5.5.3.1 Properties of Artificial Timber

- Artificial timber resembles natural wood in looks and feel.
- Artificial timber can be made in varied densities depending upon the loads expected on the structure.
- Artificial timber can be cut, curved, sawn and painted like natural wood.
- Artificial wood can be nailed or screwed like wood.
- Artificial wood is a good thermal insulator.
- Artificial wood is bio-degradable like natural wood and can be powdered and disposed off safely.

5.5.3.2 Advantages of Artificial Timber

- Artificial timber can substitute natural wood for construction purposes as it resembles it in most of its properties.
- Artificial wood has the advantage of being termite proof as the moisture content is controlled.
- Artificial wood is not inflammable. So, it can be used for doors as well as outdoor construction.
- Artificial wood can be used for heavy as well as light load carrying structures as it can be manufactured with desired density and strength.
- Artificial wood is more resistant to the action of chemicals and environmental attacks.
- Artificial wood helps to prevent deforestation and reduces the burden on natural resources.
- Artificial wood prevents pollution and minimize carbon emissions.

5.5.4 Artificial Sand

Artificial sand is manufactured by crushing rocks, stone or large aggregates into fine particles which can be used as a substitute to natural river sand. Artificial sand is also known as manufactured sand or crush sand. Artificial sand is used to act as filler between aggregates and cement, in making of concrete or mortar. It is obtained by mechanical crushing and sieving of rocks and achieving particle size less than 4.75mm. Artificial sand has higher compressive strength and cohesive properties than natural sand which makes it preferable in infrastructural construction.

5.5.4.1 Advantages of Artificial Sand

There are some significant advantages of using artificial sand which makes it popular in construction.

- Natural sand resources are formed in hundreds or thousands of years and are depleting day by day. Artificial sand can be used as a substitute to natural sand and save this natural resource.
- The use of artificial sand saves on the need of energy for mining of natural sand and also protects environment from pollution. In fact, the demolition waste from construction sites can be crushed to fine particles to get artificial sand.
- Artificial sand has high strength and better durability and hence, it is preferred to achieve high performance concrete for building construction and transportation sector. The particles of artificial sand are well graded in shape and size and are more stable compared to natural sand.
- The quality of artificial sand can be controlled as it is produced by adjustable mechanized techniques.

5.5.4.2 Artificial Sand versus Natural Sand

A comprehensive comparison between artificial and natural sand is as follows:

Table 5.1: Comparison between Artificial and Natural Sand

Property	Natural Sand	Artificial Sand
Shape of particle	Spherical	Cubical
Retention of surface moisture	7%	10%

Property	Natural Sand	Artificial Sand
Compressive Strength	Lower	Higher
Flexural Strength	Lower	Higher
Organic Impurity	More (Bones, Logs, Shells, Woods, Silt)	Negligible
Grading	Cannot be controlled	Can be controlled
Resistance against attacks	Less	More

UNIT SUMMARY

- Introduction to processed construction materials
    - Definition
    - Various types
    - Usage
- Plaster of Paris (POP) ($C_aSO_4 - 2H_2O$)
    - Properties of POP – White powder, exothermic reaction with water, sets quickly, cast in desired shape and size.
    - Types of POP – Gypsum plaster, clay plaster, lime plaster, cement plaster.
    - Uses – In dentistry, in setting fracture of bones, in making artificial ceiling, as a sealant, as a fire proofing material, to get smooth finish surface, in making of toys, show pieces.
- Paints
    - Constituents
        - Pigment
        - Resin
        - Solvent
        - Additives
    - Purpose of paint
        - Decoration
        - Protection
        - Identification
        - Sanitation
    - Characteristics of a good paint
        - Good covering capacity
        - Impervious
        - Smooth finish
        - Uniform
        - Easy to clean
        - Economical
        - Attractive
        - Wear and tear resistance.
    - Types of paints
        - Water based paint – Distemper, emulsion
        - Oil based paint
        - Cement paint
        - White wash
        - Varnish
- Industrial waste materials in construction
    - Flyash
        - Type C
        - Type F
        - Advantages of flyash
        - Uses of flyash in construction
    - Blast furnace slag
        - Advantages of blast furnace slag
        - Uses of blast furnace slag
    - Granite and marble waste
        - Advantage and uses of granite and marble waste

- Agro waste materials
  - Rice husk
  - Coir fibres
  - Bagasse
- Special processed materials
  - Geo-synthetics
    - Types – Geo-textiles, geo-grids, geo-nets, geo-membrane, geo-cement
    - Functions – Drainage, filtration, reinforcement, separation, erosion control
  - Ferrocrete
    - Properties – Low water cement ratio, high density, good bonding
    - Advantages – Thin structures, Light in weight, high early strength, easy to repair, suitable for pre-cast elements
    - Limitations – Skilled labour, less compressive strength then RCC
  - Artificial Timber
    - Properties – Varied densities can be cut/sawn/shaped, used as a replacement to natural wood
    - Advantage – Termite proof, fire proof, resistant to chemical and environmental attacks, pollution free
  - Artificial Sand
    - Properties – Cubical shape particles, high compressive and flexural strength, well graded, good resistance to attacks
    - Advantage – Quality control, energy saving, pollution control due to mining, suitable as a substitute to natural resource rive sand

EXERCISES

Multiple Choice Questions

1. Plaster of Paris has the following chemical formula:
   (a) $CaSO_3.2H_2O$ (b) $CaCl_2.2H_2O$ (c) $CaCO_3.2H_2O$ (d) $CaSO_4.2H_2O$
2. Which of the following is not an Ingredient of Paint:
   (a) Thinner (b) Pigment (c) Solvent (d) Base
3. Enamel Paints are:
   (a) Water based (b) Oil based (c) Cement based (d) None of the above
4. Which of these is an example of Water based paint:
   (a) Emulsion (b) Aluminium Paint (c) Bituminuos Paint (d) Rubber paint
5. For a glossy appearance, which of the following is used:
   (a) Distemper (b) White wash (c) Varnish (d) Emulsion
6. A suitable solvent for resinous varnish is:
   (a) Linseed oil (b) Turpentine (c) Spirit(d)All of the above

7. The ingredient of Paint which gives binding property to it is:
   (a) Filler (b) Pigment (c) Solvent (d) Base
8. Which oil helps I quick drying of the paint:
   (a) Linseed oil (b) Turpentine (c) Tung oil (d) Nut oil
9. Which of the following is NOT a problem associated with paints:
   (a) Blooming (b) Blistering (c) Fading (d) Flaking
10. __________ Varnish is also known as French varnish and is used for furniture.
   (a) Oil (b) Water (c) Acrylic (d) Spirit
11. Flyash __________ the workability of concrete.
   (a) Increases (b) Decreases (c) Does not effect (d) Can't say
12. Slag is obtained from __________ industry.
   (a) Timber (b) Thermal Power (c) Leather (d) Steel
13. Use of blast furnace slag results in __________ density.
   (a) Higher (b) Lower (c) No change (d) Can't say
14. Rice husk has high __________ content.
   (a) Silica (b) Lime
   (c) Gypsum (d) Magnesia
15. Artificial timber is better than natural timber in terms of:
   (a) Termite proof (b) Fire resistance
   (c) Compressive strength (d) All of the above

ANSWERS

1.	(d)	2.	(a)	3.	(b)	4.	(a)	5.	(c)	6.	(b)	7.	(d)	8.	(a)
9.	(a)	10.	(d)	11.	(a)	12.	(d)	13.	(b)	14.	(a)	15.	(d)		

Short Answer Type Questions

1. What are the constituents of Plaster of Paris (POP)?
2. What are the characteristics of a good paint?
3. State the effect of fly-ash on properties of concrete.
4. What are various agro-waste products used as construction materials? Discuss the use of rice-husk in construction.
5. Discuss the use of artificial sand in construction. How is artificial sand obtained?
6. What is Ferrocrete? How is it produced?
7. How is varnish different to Paint? State its uses.
8. What do you understand by cement paint?
9. State the functions of geo-synthetics.
10. Explain the objectives of varnishing.

Long Answer Type Questions

11. What are geo-synthetics? Discuss various types of geo-synthetics, mentioning their area of application.
12. What is blast furnace slag? How is it obtained? State the advantages and uses of using slag in construction.
13. Explain the use of fly ash in concrete stating its advantages and applications.
14. What are the constituents of paint? State the types and functions of paints.
15. Compare natural timber with artificial timber in terms of its properties and uses.

PRACTICALS

Experiment-1

Aim: Apply two or more coats of selected paint on the prepared base of a given wall surface for the area of 1 m × 1 m using suitable brush/rollers adopting safe practices.

Apparatus/Materials Required: Paint Scraper, Sand paper, Brushes, Roller

Theory: Paints are applied on wall surface to provide a smooth, attractive and easy to clean surface. The walls or any surface to be painted in first prepared with a base before applying any paint on it. The paints are either applied on a new wall surface or the wall surfaces are repainted, as per requirement of cleanliness or change in its look.

Procedure:

1. **Preparation of space:** The space where wall surface is to be painted needs to be protected from splash of pint. For it, more all the furniture or appliance to the middle of the room to prevent any obstacle in painting covers all the items with plastic cover.
2. **Removals of old paint:** The wall surface to be painted needs to be scrapped off of any previous applied paint. The old paint might be seen flaking off or curling up. Paint scraper is used for rubbing and scrapping off the old paint. This should be done with at most care and efficiently to get a uniform layer of paint.
3. **Sand the Surface:** The scraped paints of wall are sometimes uneven and hence, the after scrapping, the wall surface is subbed with coarse sand paper to get a smooth surface. Painting is done on uneven surface may have edges and ridges. A vigorous rubbing with sand paper is desired to get a smooth and even surface after painting.
4. **Filling the gapes and cracks:** Before applying paint on the subbed wall surface, it is inspected for any gaps or cracks. These cracks are filled with sealants and fillers (like putty) to get perfect smooth surface on painting.
5. **Cleaning the wall surface:** After filling the creaks, the wall surface is cleaned with a soap water or cleaner to remove any dirt, grease or grime. This step is done to ensure proper adhering of the new paint over the surface.

6. **Priming the wall:** An even coat of primer or sealer is applied on the wall surface to get a smooth finish. The primer coat makes the painting cost efficient as applying a coat of primer is economical than applying an additional coat of paint. Primer coat covers the old colour paint, if any and gives a uniform finish of new paint.
7. **Apply the paint:** Apply a coat of new paint either by brush or roller on the prepared wall surface. At least two coats of new paint are recommended to get a good finish. Each coat of paint is applied only after the previous coat is dried properly.

Precautions:

1. Make sure that no old paint is left over the surface to be painted.
2. Ensure to remove any dirt or grease as it may causes improper adherence of the paint to the surface.
3. Allow each coat of paint to dry thoroughly before applying the next layer.

Experiment-2

Aim: Prepare the cement mortar of proportion 1:3 or 1:6 using artificial sand as a special processed construction material.

Apparatus/Materials Required: Raw materials i.e., Cement, Artificial Sand, Water, Shovel, Mixing Tray, Weighing Balance and Measuring Cylinder.

Theory: Traditionally, mortar is defined as a homogenous mixture of cement, sand and water. Cement acts as a binding material and provides strength to the mortar. Fine aggregate i.e., sand, provider strength, resistance to shrinkage and cracking in mortar. It also provides bulk to mortar and makes it economical. Water is added to activate the hydration process in mortar to achieve the desired strength. Sand can be replaced with many suitable materials like artificial sand/flyash/granite or marble polishing waste powder. Mortar is provided to get an impervious smooth surface in the form of plaster on walls. It is also used to bind the masonry units like bricks, stones or blocks. Mortar acts as a filler material also in masonry works. It further helps in controlling/reducing shrinkage cracks.

Procedure:

1. **Selection of raw materials:** Choose right raw materials for the desired strength of the mortar and its applicability. The Protland cement is generally suitable for preparing mortar. Lime may also be used as a binding material. The filler material may be sand or artificial sand or flyash or granite powder or marble waste powder. Selection of suitable raw materials is an important decision depending on the type of construction and the purpose.
2. **Proportioning the mortar:** The proportioning means deciding the relative quantity of different raw materials to be mixed to achieve a good mortar for the desired application. Decide the proportions of raw material on the basis of following recommendations:
   (a) For masonry work – 1:3 to 1:6
   (b) For reinforced brick work – 1:2 to 1:3
   (c) For external plastering work – 1:4

(d) For internal plastering work – 1:5 to 1:6

(e) For flooring work – 1:4 to 1:8

(f) For pointing work – 1:1 to 1:3

3. **Mixing of ingredients:** Firstly, dry mix the cement and sand/artificial sand/ flyash/marble powder properly in right proportions. Add water in mix gradually using a shovel. Cement mortar can either be mixed manually for routine projects or mechanically using mixers for large projects.
4. **Curing the mortar:** Cement gains strength with hydration. It ensures continues hydration, mortar is kept wet. Curing is the process of keeping mortar wet. Curing of mortar is done by spraying water on it. Generally, 7 days curing of mortar is recommended.

Precautions:

1. The raw materials should not have any lumps and should be free impurities.
2. The materials should be weighted accurately for right proportioning.
3. The materials should be dry mixed first before adding any water.
4. Sufficient and proper curing should be ensured to achieve the desired strength properties of the mortar.

KNOW MORE

Paints, as categorized, are either water based or oil based. Distempers are the most common type of water based paints. However, it is noteworthy to study the features of distempers. On drying, distemper shrinks and cracks. Moreover, the coating of distemper is thick and brittle and hence flakes out with the passage of time. Also, the coating provided by distemper is porous and allows water to pass through it. The use of distemper is NOT recommended in damp locations like kitchens, bathrooms etc. A superior type of paint category is oil based, also known as enamels which give a better strength and durability. Cement paints, a special type of water based paint, is preferred over exterior surfaces as the last protective coating. There are few other special types of paints like aluminium paint (for iron and steel surfaces), Asbestos paint (for repair works to control leakages), Bituminous or tar paints (for under water works), cellular paints (a superior paint based on evaporation of solvent for hardening), plastic paint (a superior paint with plastics as the base), silicate paint (requires no priming coat) and synthetic rubber paint (excellent paint with synthetic resins as base).

Varnishes have similar objectives as that of paints like to provide a protective and attractive film on the surface. The characteristics of a good varnish are also similar to that of a good paint but it should also give a glossy look. The ingredients of a varnish are resin, solvent and drier. English copal varnish is the best varnish available. Multiple coats of varnish are applied on a surface to get the desired glow.

The utilization of industrial waste products in construction has considerable sustainability benefits. The safe disposal of these by products is difficult and requires huge land for fills. The disposal in landfills or in water bodies poses a serious threat in terms of land and water pollution. It effects the environment adversely. Moreover, the natural construction materials are limited in resources or the artificial materials are very energy intensive in their production. Cement itself requires a lot of energy for production and releases huge amount of CO_2, responsible for global warming. The use of industrial by-products like flyash and blast furnace slag in construction reduces the demand of conventional construction materials and give an eco-friendly approach to construction activities. The government of India is also promoting the use of industrial wastes in construction through its codes and creating awareness on the topic in masses.

A lot of agro-waste materials are gaining popularity as building material. One such material is coir fibre which is a natural fibre extracted from the husk of the coconut. The coir fibre is the thickest and most resistant of all natural fibres. The low rate of decomposition and its water resistant properties are making coir fibres popular in construction. Coir fibres have further the advantage of lower thermal conductivity and hence are used as an insulating material during construction.

Bagasse is another agro waste material used as supplementary cementing material used in construction. Sugarcane bagasse ash, which is obtained by burning agricultural waste bagasse, can be used as a cement replacement. This can reduce the carbon emissions from cement industry and have high economic and environmental benefits. Bagasse ash improves the fresh, strength and durability properties of concrete. It is noteworthy here that bagasse ash cannot be used as a stabilizer alone and needs to be mixed with cement to act as a soil stabilizer.

Most of the agro based waste materials are fibrous in nature and can be utilised in concrete as a replacement to conventional ingredients or as an additive to improve the physical, mechanical, strength and durability properties of concrete. When agro based wastes are burnt to produce ash it gains some pozzolanic properties which can be utilized to act as binder and improve the properties further. Agro based wastes have been utilized as a filler in concrete. Alkali-activated systems have also been tried to produce geopolymer concrete using agro wastes. The promising results of latest research have shown the potential of utilization of agro based wastes in construction industry.

Ferrocrete is a new construction material which originated from France in 19th century for manufacturing of a rowing boat. Later, it was used in construction activities in Italy in 20th century. Ferrocement and ferrocrete are same. Though ferrocrete is becoming popular construction material now a days, its long term performance is yet to be ascertained.

Artificial timber is one of the most sustainable innovations in the field of construction materials. Though the material is still under research stage, it has shown many advantages over the use of natural wood. The raw materials for artificial timber can be, not only fly ash, bitumen or bottom ash but many more waste materials may be studied to develop artificial timber. The manufacturing of artificial timber is not an expensive procedure, nor does it require any special machinery or skill set. Moreover, it provides a solution for a cleaner and sustainable construction industry.

Artificial sand is another move towards sustainable construction. Artificial sand is obtained by crushing rocks, aggregates or demolition waste under controlled conditions by mechanical means ensuring desired quality and grading. The demand of artificial sand is rising over natural sand in modern era of construction due to its better performance in terms of strength and durability. Artificial sand has the advantage of energy saving & pollution control besides safeguarding the exploitation of natural resource river sand.

REFERENCES AND SUGGESTED READINGS

- Rajput, R.K, Engineering Materials, S. Chand and Co., New Delhi.
- S. K. Sharma, Civil Engineering Construction Materials, Khanna Publishing House, Delhi

CO AND PO ATTAINMENT TABLE

Course outcomes (COs) for this course can be mapped with the programme outcomes (POs) after the completion of the course and a correlation can be made for the attainment of POs to analyze the gap. After proper analysis of the gap in the attainment of POs necessary measures can be taken to overcome the gaps.

Table for CO and PO attainment

Course Outcomes	Attainment of Programme Outcomes (1- Weak Correlation; 2- Medium correlation; 3- Strong Correlation)						
	PO-1	PO-2	PO-3	PO-4	PO-5	PO-6	PO-7
CO-1							
CO-2							
CO-3							
CO-4							
CO-5							

The data filled in the above table can be used for gap analysis.

REFERENCES FOR FURTHER LEARNING

1. Ghose, D. N., Construction Materials, Tata McGraw Hill, New Delhi.
2. S.K. Sharma, Civil Engineering Construction Materials, Khanna Publishing House, Delhi.
3. Varghese, P.C. , Building Materials, PHI learning, New Delhi.
4. Rangwala, S.C., Engineering Materials, Charator publisher, Ahemdabad.
5. Somayaji, Shan, Civil Engineering Materials, Pearson education, New Delhi.
6. Rajput, R.K, Engineering Materials, S. Chand and Co., New Delhi.
7. Sood H., Laboratory Manual on Testing of Engineering Materials, New Age Publishers, New Delhi.
8. Sharma C. P., Engineering Materials, PHI Learning, New Delhi.
9. Duggal, S. K, Building Materials, New International, New Delhi.
10. Arora S.P and Bindra S.P., Building Construction, Dhanpat Rai Publication, New Delhi.

INDEX